T0144495

Graph Databases
Applications on Social Media Analytics and Smart Cities

Editor

Christos Tjortjis

Dean, School of Science and Technology
International Hellenic University
Greece

CRC Press
Taylor & Francis Group
Boca Raton London New York

CRC Press is an imprint of the
Taylor & Francis Group, an **informa** business

A SCIENCE PUBLISHERS BOOK

First edition published 2024
by CRC Press
2385 NW Executive Center Drive, Suite 320, Boca Raton FL 33431

and by CRC Press
4 Park Square, Milton Park, Abingdon, Oxon, OX14 4RN

© 2024 Taylor & Francis Group, LLC

CRC Press is an imprint of Taylor & Francis Group, LLC

Reasonable efforts have been made to publish reliable data and information, but the author and publisher cannot assume responsibility for the validity of all materials or the consequences of their use. The authors and publishers have attempted to trace the copyright holders of all material reproduced in this publication and apologize to copyright holders if permission to publish in this form has not been obtained. If any copyright material has not been acknowledged please write and let us know so we may rectify in any future reprint.

Except as permitted under U.S. Copyright Law, no part of this book may be reprinted, reproduced, transmitted, or utilized in any form by any electronic, mechanical, or other means, now known or hereafter invented, including photocopying, microfilming, and recording, or in any information storage or retrieval system, without written permission from the publishers.

For permission to photocopy or use material electronically from this work, access www.copyright.com or contact the Copyright Clearance Center, Inc. (CCC), 222 Rosewood Drive, Danvers, MA 01923, 978-750-8400. For works that are not available on CCC please contact mpkbookspermissions@tandf.co.uk

Trademark notice: Product or corporate names may be trademarks or registered trademarks and are used only for identification and explanation without intent to infringe.

Library of Congress Cataloging-in-Publication Data (applied for)

ISBN: 978-1-032-02478-3 (hbk)
ISBN: 978-1-032-02479-0 (pbk)
ISBN: 978-1-003-18353-2 (ebk)

DOI: 10.1201/9781003183532

Typeset in Times New Roman
by Innovative Processors

Preface

The idea for the book came about after fruitful discussions with members of the Data Mining and Analytics research group, stemming out of frustrations using conventional database management systems for our research in Data Mining, Social Media Analytics, and Smart cities, as well as aspirations to enhance the utilisation of complex heterogeneous big data for our day-to-day research. The need was confirmed after discussions with colleagues across the globe, as well as surveying the state of the art, so we happily embarked on the challenge to put together a high-quality collection of chapters complementary but coherent, telling the story of the ever-increasing rate of graph database usage, especially in the context of social media and smart cities. Meanwhile, our planet was taken aback by the new pandemic storm. Plans were disrupted, priorities changed, and attention was diverted towards more pressing matters. Yet, the notion of the need for social media analytics coupled with smart city applications including healthcare and facilitated by graph databases emerged even stronger.

The editor is grateful to all authors who weathered the storm for their contributions and to the editorial team for their support throughout the compilation of this book. I hope that the selected chapters offer a firm foundation, but also new knowledge and ideas for readers to understand, use and improve applications of graph databases in the areas of Social Media Analytics, Smart Cities, and beyond.

September 2022 **Christos Tjortjis**

Contents

Introduction

Christos Tjortjis
Dean of the School of Science and Technology, International Hellenic University
email: c.tjortjis@ihu.edu.gr

With Facebook having 2.9 billion active users, YouTube following with 2.5 billion, Instagram with 1.5 billion, TikTok with 1 billion, and Twitter with 430 million, the amount of data published daily is excessive. In 2022 it was estimated that 500 million tweets were published daily, 1 billion get posted daily across Facebook apps, there were 17 billion posts with location tracking on Facebook, and 350 million uploaded daily, with the accumulated number of uploaded photos reaching 350 billion. Every minute 500 hours of new video content is uploaded on YouTube, meaning that 82, 2 years of video content is uploaded daily. On Instagram, 95 million photos and videos are uploaded daily. The importance of gathering such rich data, often called "the digital gold rush", processing it, and retrieving information are vital.

Graph databases have gained increasing popularity recently, disrupting areas traditionally dominated by conventional *relational*, *SQL*-based databases, as well as domains requiring the extra capabilities afforded by graphs. This book is a timely effort to capture the state of the art of Graph databases and their applications in domains, such as *Social Media analysis* and *Smart Cities*.

This practical book aims at combining various advanced tools, technologies, and techniques to aid understanding and better utilizing the power of Social Media Analytics, *Data Mining,* and Graph Databases. The book strives to support students, researchers, developers, and simple users involved with *Data Science* and Graph Databases to master the notions, concepts, techniques, and tools necessary to extract data from social media or smart cities that facilitate information acquisition, management, and prediction.

The contents of the book guide the interested reader into a tour starting with a detailed comparison of relational SQL Databases with NoSQL and Graph Databases, reviewing their popularity, with a focus on Neo4j.

Chapter 1 details the characteristics and reviews the pros and cons of relational and NoSQL databases assessing and explaining the increasing popularity of the latter, in particular when it comes to Neo4j. The chapter includes a categorization of NoSQL Database Management Systems (DBMS) into i) Column, ii) Document, iii) Key-value, iv) Graph and v) TimeSeries. Neo4j Use Cases and related scientific research are detailed, and the chapter concludes with an insightful discussion. It is essential reading for any reader who is not familiar with the related concepts before engaging with the following chapters.

Next, two surveys review the state of play regarding graph databases and social media data. The former emphasises analytics from a *link prediction* perspective and the latter focuses on *knowledge extraction* from social media data stored in Neo4j. Graph databases can manage highly connected data originating from social media, as they are suitable for storing, searching, and retrieving data that are rich in relationships.

Chapter 2 reviews the literature for graph databases and software libraries suitable for performing common social network analytic tasks. It proposes a *taxonomy* of graph database approaches for social network analytics based on the available algorithms and the provided means of storing, importing, exporting, and querying data, as well as the ability to deal with big social graphs, and the corresponding CPU and memory usage. Various graph technologies are evaluated by experiments related to the link prediction problem on datasets of diverse sizes.

Chapter 3 introduces novel capabilities for knowledge extraction by surveying Neo4j usage for social media. It highlights the importance of transitioning from SQL to NoSQL databases and proposes a categorization of Neo4j use cases in Social Media. The relevant literature is reviewed including various domains, such as Recommendation systems, marketing, learning applications, Healthcare analytics, Influence detection, and Fake news.

The theme is further developed by two more chapters: one elaborating on *combining multiple social networks* on a single graph, and another on YouTube child influencers and the relevant *community detection*.

Chapter 4 makes the case for combining multiple Social Networks into a Single Graph, since one user maintains several accounts across a variety of social media platforms. This combination brings forward the potential for improved recommendations enhancing user experience. It studies actual data from thousands of users on nine social networks (Twitter, Instagram, Flickr, Meetup, LinkedIn, Pinterest, Reddit, Foursquare, and YouTube). *Node similarity* methods were developed, and node matching success was increased. In addition, a new *alignment method* for multiple social networks is proposed. Success rates are measured and a broad user profile covering more than one social network is created.

Chapter 5 investigates data collection and analysis about child Influencers and their follower communities on YouTube to detect *overlapping communities* and understand the socioeconomic impact of child influencers in different cultures. It presents an approach to data collection, and storage using the graph database ArangoDB, and analysis with overlapping community detection algorithms, such as SLPA, CliZZ, and LEMON. With the open source WebOCD framework, community detection revealed that communities form around child influencer channels with similar topics, and that there is a potential divide between family channel communities and singular child influencer channel communities. The network collected contains 72,577 channels and 2,025,879 edges with 388 confirmed child influencers. The collection scripts, the software, and the data set in the database are available freely for further use in education and research.

The smart city theme is investigated in three chapters. First a comprehensive literature survey on using graph databases to manage smart city *linked data*. Two case studies follow, one emphasising *energy load forecasting* using graph databases and *Machine Learning*, and another on *digital health* applications which utilise a Graph-Based data model.

Chapter 6 provides a detailed literature survey integrating the concepts of Smart City Linked Data with Graph Databases and social media. Based on the concept of a smart city as a complex linked system producing vast amounts of data, and carrying many connections, it capitalises on the opportunities for efficient organization and management of such complex networks provided by Graph databases, given their high performance, flexibility, and agility. The insights gained through a detailed and critical review and synthesis of the related work show that graph databases are suitable for all layers of smart city applications. These relate to social systems including people, commerce, culture, and policies, posing as user-generated in social media. Graph databases are an efficient tool for managing the high density and interconnectivity that characterizes smart cities.

Chapter 7 ventures further into the domain of smart cities focusing on the case of *Energy Load Forecasting (ELF)* using Neo4j, the leading NoSQL Graph database, and Machine Learning. It proposes and evaluates a method for integrating multiple approaches for executing ELF tests on historical building data. The experiments produce data resolution for 15 minutes as one step ahead of the time series forecast and reveal accuracy comparisons. The chapter provides guidelines for developing correct insights for energy demand predictions and proposes useful extensions for future work.

Finally, **Chapter 8** concludes with an interesting Graph-Based Data Model for Digital Health Applications in the context of smart cities. A key challenge for modern smart cities is the generation of large volumes of heterogeneous data to be integrated and managed to support the discovery of complex relationships in domains like healthcare. This chapter proposes a methodology for transferring a relational to a graph database by mapping the relational schema to a graph schema. To this end, a relational schema graph is constructed for the relational database and transformed in several steps. The approach is demonstrated in the example of a graph-based medical information system using a dashboard on top of a Neo4j database system to visualize, explore and analyse the stored data.

From Relational to NoSQL Databases – Comparison and Popularity Graph Databases and the Neo4j Use Cases

Dimitrios Rousidis [0000-0003-0632-9731] and
Paraskevas Koukaras [0000-0002-1183-9878]

The Data Mining and Analytics Research Group, School of Science and Technology, International Hellenic University, GR-570 01 Thermi, Thessaloniki, Greece
e-mail: d.rousidis@ihu.edu.gr, p.koukaras@ihu.edu.gr

In this chapter an in-depth comparison between Relational Databases (RD) and NoSQL Databases is performed. There is a recent trend for the IT community and enterprises to increasingly rely on the NoSQL Databases. This chapter briefly introduces the main types of NoSQL Databases. It also investigates this trend by discussing the disadvantages of RD and the benefits of NoSQL Databases. The interchanges of the popularity in the past 10 years of both types of Databases are also depicted. Finally, the most important Graph Databases are discussed and the most popular one, namely Neo4j, is analyzed.

1.1   Introduction

In Codd We Trust. Published on March 6th, 1972, the paper with the title "Relational Completeness of Data Base Sublanguages" [1] written by Edgar Frank "Ted" Codd (19 August 1923–18 April 2003), an Oxford-educated mathematician working for IBM, was one of the most seminal and ground-breaking IT publications of the 20th century. The abstract of the publication starts with "In the near future, we can expect a great variety of languages to be proposed for interrogating and updating data bases. This paper attempts to provide a theoretical basis which may be used to determine how complete a selection capability is provided in a proposed data sublanguage independently of any host language in which the sublanguage may be embedded", whilst the last section of the paper with the heading "Calculus Versus Algebra" begins with "A query language (or other data sublanguage) which is claimed to be general purpose should be at least relationally complete in the sense defined in this paper. Both the algebra and calculus described herein provide a foundation for designing relationally complete query languages without resorting to programming loops or any other form of branched execution – an important consideration when interrogating a data base from a terminal". The paper, which proved to be prophetic, proposed a relational data model for massive, shared data banks and established a relational model based on the mathematical set theory and laid the foundation for a world created and organized by relations, tuples, attributes and relationships. In the 70's Codd and CJ Date developed the relational model, 1979 was the year that Oracle developed the first commercial Relational Database Management System (RDBMS) and since, after 40 plus years is still the predominate database model.

Since then, the world has entered the Big Data era and according to an article by Seed Scientific titled "How much data is created every day? – 27 staggering facts" the total volume of data in the world was predicted to reach 44 zettabytes at the start of 2020, with more than 2.5 quintillion bytes produced each day, and Google, Facebook, Microsoft, and Amazon storing at least 1,200 petabytes of data [2]. These numbers are expected to get even bigger as Internet of Things (IoT) connected devices are projected to reach more than 75 billion and globally, the volume of data created each day is predicted to reach 463 exabytes by 2025 [3]. The constant rise of use of Social Media (SM) [14, 15] and Cloud Computing provide many useful mechanisms in our digital world [16], but at the same time this massive production of data boomed the volume and forced the IT industry to search for new, more powerful, flexible and able to cope with enormous datasets Database Management Systems (DBMS). Therefore, NoSQL, originally referring to "non-SQL" but nowadays referring to "Not-Only SQL", a term introduced by C. Strozzi in 1998 [4], provides a way for enhancing the features of standard RDBMS.

The aim of this chapter is to articulate on the advantages of NoSQL DBs, and also the prospects and potential that can be realized by incorporating Machine Learning (ML) methods. It is foreseen to provide practitioners and academics with sufficient reasoning for the need of NoSQL DBs, what types can be used depending on multiple occasions, and what mining tasks can be conducted with them. The rest of the chapter is structured as follows: Section 2 presents the characteristics, the advantages and disadvantages of both RDs and NoSQL DBs, the categorization of NoSQL DBs. Section 3 analyzes the popularity of the NoSQL DBs, based on the ranking of the db-engines.com website. The next section focuses on Graph DBMS and the leading DBMS, Neo4j and its most important use cases are presented. The chapter concludes with the most important findings of this research.

1.2   NoSQL Databases

1.2.1   Characteristics of NoSQL Databases

In comparison to the Relational Databases' ACID (Atomicity, Consistency, Isolation, Durability) concept, NoSQL is built on the BASE (Basically Available, Soft State, and Eventually Consistent) concept. Its key benefit is the simplicity with which it can store, handle, manipulate, and retrieve massive amounts of data, making it perfect for data-intensive internet applications [5] and giving several functional advantages and data mining capabilities [6]. The following are the primary features of NoSQL DBs:

1. Non-relational: They do not completely support relational DB capabilities like as joins, for example.
2. No Schema (they do not have a fixed data structure).
3. Data are replicated across numerous nodes, making it fault-tolerant.
4. Horizontal scalability (connecting multiple hardware or software entities so that they work as a single logical unit).
5. Since they are open source, they are inexpensive and simple to install.
6. Outstanding write-read-remove-get performance.
7. Stable consistency (all users see the same data).
8. High availability (every user has at least one copy of the desired data).
9. Partition-tolerant (the total system keeps its characteristics, even when being deployed on different servers, transparently to the client).

1.2.2   Drawbacks and Advantages

According to [7] the main uses of NoSQL in industry are:

1. Session store (which manages session data).
2. User profile store (which enables online transactions and creates a user-friendly environment).
3. Content and metadata store (building a data and metadata warehouse, enabling the storage of different types of data).
4. Mobile applications.
5. IoT (assisting the concurrent expansion, access, and manipulation of data from billions of devices).
6. Third-party aggregation (with the ease of managing massive amounts of data, allowing access by third-party organizations).
7. E-commerce (storing and handling enormous volumes of data).
8. Social gaming.
9. Ad-targeting (enabling tracking user details quickly).

From Table 1.1, where the main advantages and disadvantages of NoSQL have been gathered and summarily explained, it is obvious that the benefits of NoSQL prevail over its few drawbacks, constituting NoSQL very effective, providing better performance, and offering a cost-effective way of creating, collecting, manipulating, querying, sharing and visualizing data.

Table 1.1. Advantages/Disadvantages of NoSQL over SQL [8]

Advantages	Explanation
Non-relational	Not tuple-based – no joins and other RD features
Schema-less	Not strict/fixed structure
Data are replicated to multiple nodes and can be partitioned	Down nodes are simply replaced, and there is no single point of failure
Horizontally scalable	Cheap, simple to set up (open-source), vast write performance, and quick key-value access
Provides a wide range of data models	Supports new/modern datatypes and models
Database administrators are not required	Not direct management and supervision required
Less hardware failures	NoSQL DBaaS providers such as Riak and Cassandra are designed to deal with equipment failures
Faster, more efficient, and flexible	Simple, scalable, efficient, multifunctional
Has evolved at a very high pace	Fast growth by IT leading companies
Less time writing queries	More time comprehending answers – more condensed and functional queries
Less time debugging queries	More time spent developing the next piece of code, elevated overall code quality
Code is easier to read	Faster ramp-up for new project members, improved maintainability and troubleshooting
The growing of big data - in that high data velocity, data variety, data volume, and data complexity	Data is constantly available. True spatial transparency. Transactional capabilities for the modern era. Data architecture that is adaptable. High-performance architecture with a high level of intelligence
It has huge volumes of fast changing structured, semi-structured, and unstructured data that are generated by users	Quick schema iteration, agile sprints and frequently code pushes quickly. Support for object-oriented programming languages that are simple to comprehend and use in a short amount of time. NoSQL is a globally distributed scale-out architecture that is not costly and monolithic. Agnostic to schema, scalability, speed, and high availability. Handles enormous amounts of data with ease and at a cheaper cost
Disadvantages	**Explanation**
Immature	Need additional time to acquire the RDs' consistency, sustainability and maturity
No standard query language	Compared to RD's SQL
Some NoSQL DBs are not ACID compliant	Atomicity, consistency, isolation, durability offers stability
No standard interface. Maintenance is difficult	Some DBs do not offer GUI yet

1.2.3   Categorization

According to the bibliography, there are five main/dominant categories for NoSQL DBMS: (1) Column, (2) Document, (3) Key-value, (4) Graph and (5) Time Series. In "NoSQL Databases List by Hosting Data" [9] it is mentioned that there are 15 categories; the five aforementioned ones and ten more that were characterized as Soft NoSQL Systems (6 to 15): (6) Multimodel DB, (7) Multivalue DB, (8) Multidimensional DB, (9) Event Sourcing, (10) XML DB, (11) Grid & Cloud DB Solutions, (12) Object DB, (13) Scientific and Specialized DBs, (14) Other NoSQL related DB, and finally (15) Unresolved and uncategorized [10], whilst the db-engines.com website has 15 categories in total, too, to be discussed in the following section.

Next, the five most popular categories are analyzed as described at the db-engines website:

(1) Key-value stores[1]: Key-Value which are based on Amazon's Dynamo paper [11] and "they are considered to be the simplest NoSQL DBMS since they can only store pairs of keys and values, as well as retrieve values when a key is known".

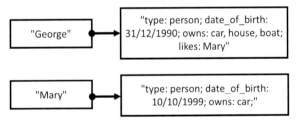

Figure 1.1. Key-value data.

(2) Wide column stores[2]: They were first introduced at Google's BigTable paper [12], and they are "also called extensible record stores, store data in records with an ability to hold very large numbers of dynamic columns. Since the column names as well as the record keys are not fixed, and since a record can have billions of columns, wide column stores can be seen as two-dimensional key-value stores".

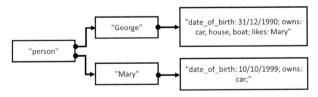

Figure 1.2. Wide column data.

---

[1] https://db-engines.com/en/article/Key-value+Stores
[2] https://db-engines.com/en/article/Wide+Column+Stores

(3) Graph databases[3]: Also called graph-oriented DBMS, they represent data in graph structures as nodes and edges, which are relationships between nodes. "They allow easy processing of data in that form, and simple calculation of specific properties of the graph, such as the number of steps needed to get from one node to another node".

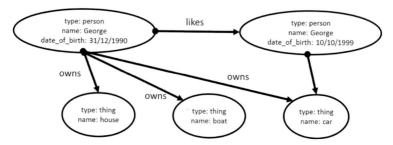

Figure 1.3. Graph databases data.

(4) Document stores[4]: "Document stores, also called document-oriented database systems, are characterized by their schema-free organization of data". That is, the columns can have more than one value (arrays), the records can have a nested configuration, and the configuration of the records and the types of data are not obliged to be uniform. Thus, different records can have different columns, and the values of individual columns may be different for each record.

persons	

{
name: "George",
date_of birth: "31/12/1999",
owns: ["car", "house", "boat",]
like: "Mary"
}
	{
	name: "Mary"
	date_of_birth: "10/10/1990",
	owns: "car"
	}

Figure 1.4. Document based data.

(v) Time series[5]: "A Time Series DBMS is a database management system that is optimized for handling time series data: each entry is associated with a timestamp. Time Series DBMS are designed to efficiently collect, store and query various time series with high transaction volumes. Although time series data can be managed with other categories of DBMS (from key-value stores to relational systems), the specific challenges often require specialized systems".

---

[3] https://db-engines.com/en/article/Graph+DBMS
[4] https://db-engines.com/en/article/Document+Stores
[5] https://db-engines.com/en/article/Time+Series+DBMS

Measurement Time	System Threads	System Processes
2021/07/01 00:00	1,078	65
2021/07/01 01:00	1,119	66
2021/07/01 02:00	654	39
...		
2021/07/01 21:00	655	33
2021/07/01 22:00	1,251	72
2021/07/01 23:00	1,975	99

Figure 1.5. Time series data.

1.3   Popularity

In this section the popularity of the Relational and the NoSQL databases is discussed, based on the ranking of an excellent website (db-engines.com). The creators of db-engines.com have accumulated hundreds of databases and they have developed a methodology where all these databases are ranked based on their popularity

1.3.1   DB engines

The creators of db-engines claim that the platform "is an initiative to collect and present information on database management systems (DBMS). In addition to established relational DBMS, systems and concepts of the growing NoSQL area are emphasized. The DB-Engines Ranking is a list of DBMS ranked by their current popularity. The list is updated monthly. The most important properties of numerous systems are shown in the overview of database management systems". Each system's attributes may be examined by the user, and they can be compared side by side. This topic's words and concepts are discussed in the database encyclopedia. Recent DB-Engines news, citations and major events are also highlighted on the website.

1.3.2   Methodology

The platform's creators established a system for computing DBMS scores termed 'DB-Engines Ranking', which is a list of DBMS rated by their current popularity. They use the following parameters to assess a system's popularity[6]:

- Number of mentions of the system on websites, assessed by the number of results in Google and Bing search engine inquiries. To count only relevant results, they search for the system name followed by the phrase database, such as 'Oracle' and 'database'. General interest in the system, measured by the number of queries in Google Trends..
- Frequency of technical discussions about the system. They utilize the number of similar queries and interested users on the well-known IT-related Q&A sites Stack Overflow and DBA Stack Exchange.

---

[6] https://db-engines.com/en/ranking_definition

- Number of job offers, in which the system is referenced and where they utilize the number of offers on the top job search engines Indeed and Simply Hired.
- Number of profiles in professional networks, where the system is referenced, as determined by data from the most prominent professional network LinkedIn.
- Relevance in social networks, where they calculate the number of tweets, in which the system is cited.

They calculate the popularity value of a system "by standardizing and averaging of the individual parameters. These mathematical transformations are made in a way so that the distance of the individual systems is preserved. That means, when system A has twice as large a value in the DB-Engines Ranking as system B, then it is twice as popular when averaged over the individual evaluation criteria".

To remove the impact induced by changing amounts of the data sources themselves, the popularity score is a relative number that should only be understood in relation to other systems. However, the DB-Engines Ranking does not take into account the number of systems installed or their application in IT systems.

On the website there is information on 432 database systems (https://db-engines.com/en/systems), which are examined and divided in 15 categories. However, just 381 databases are ranked in accordance with the aforementioned methodology. The number and percentage of databases per category is shown in Table 1.2.

Table 1.2. Number of DBMS by category

Categories	Number of databases	Percentage (%)
Relational DBMS	152	39.90
Key-value stores	64	16.80
Document stores	53	13.91
Time series DBMS	39	10.24
Graph DBMS	36	9.45
Object oriented DBMS	21	5.51
Search engines	21	5.51
Wide column stores	21	5.51
RDF stores	20	5.25
Multivalue DBMS	11	2.89
Native XML DBMS	7	1.84
Spatial DBMS	5	1.31
Event stores	3	0.79
Content stores	2	0.52
Navigational DBMS	2	0.52

It is evident that the sum of the total number of databases exceeds 381, since there are databases that belong to more than one category.

1.3.3 Measuring Popularity

As per the DB-Engines ranking, "an initiative to collect and present information on DBMS" that ranks (updated monthly) DB management systems based on their popularity, NoSQL DBs are steadily on the rise, whereas relational DBs, while still at the top, either remain intact or show a minor declining trend. The DB-engines technique for determining a system's popularity is based on the following six parameters: (1) Number of mentions of the system on the website, (2) General interest in the system, (3) Frequency of technical debates about the system, (4) Number of job offers mentioning the system, (5) Number of profiles in professional networks mentioning the system, and (6) Relevance in social networks.

The top 20 most popular DBMSs, with their score are presented in Table 1.3, which also includes the change of their position since the previous month and the previous year (2020).

Table 1.3. Databases popularity (December 2021)

Rank			DBMS	Database model	Score
Dec 2021	Nov 2021	Dec 2020			Dec 2021
1	1	1	Oracle	Relational, Multi-model	1,281.74
2	2	2	MySQL	Relational, Multi-model	1,206.04
3	3	3	Microsoft SQL Server	Relational, Multi-model	954.02
4	4	4	PostgreSQL	Relational, Multi-model	608.21
5	5	5	MongoDB	Document, Multi-model	484.67
6	6	7	Redis	Key-value, Multi-model	173.54
7	7	6	IBM Db2	Relational, Multi-model	167.18
8	8	8	Elasticsearch	Search engine, Multi-model	157.72
9	9	9	SQLite	Relational	128.68
10	11	11	Microsoft Access	Relational	125.99
11	10	10.	Cassandra	Wide column	119.2
12	12	12	MariaDB	Relational, Multi-model	104.36
13	13	13	Splunk	Search engine	94.32
14	15	16	Microsoft Azure SQL Database	Relational, Multi-model	83.25
15	14	15	Hive	Relational	81.93
16	16	17	Amazon DynamoDB	Multi-model	77.63
17	18	41	Snowflake	Relational	71.03
18	17	14	Teradata	Relational, Multi-model	70.29
19	19	19	Neo4j	Graph	58.03
20	22	21	Solr	Search engine, Multi-model	57.72

It is obvious that the popularity of the relational DB is not disputed, ranking in the first four positions; however, NoSQL DBs are on the rise. In the 20 most popular DBMS systems there are 12 (60%) categorized as Relational Databases, 3 (15%) categorized as Search Engines and 1 (5%) are Key-Value, Document, Graph and Wide-Column. Amazon DynamoDB has a dual-equal categorization as Document and Key-Value DBMS and in the db-engines ranking is referred as a Multi-model DBMS. From the 20 most popular DBMS, 13 in total are characterized as multi-model databases, along with their primary/main categorization.

MongoDB is the most popular Document DB, at the 5th position, the most popular Key-Value DB is Redis (at the 6th position), Cassandra is 10th overall and it is the most popular Wide Column DB, Neo4j is leading the Graph DBs at the 19th position. InfluxDB is the most popular Time-Series DB at the 29th place.

A comparison between the popularity of the Top-3 RDBMS and the non-relational DBMS from the Top-20, along with the most popular Time-Series DBMS is depicted in Table 1.4. Table 1.4 demonstrates how much more popular were Oracle, MySQL and MS SQL Server DBMS, back in December 2016 compared to today (December 2021).

According to Table 1.4, the gap in popularity between RDBMS and the non-RDBMS has been reducing as the top-3 RDBMS are becoming less popular, whereas the non-RDBMS are becoming more popular every year.

Table 1.4. Top 3 RDBMS vs. NoSQL DBMS popularity

DBMS	Database model	Oracle		MySQL		Microsoft SQL server	
		2016	2021	2016	2021	2016	2021
MongoDB	Document, Multi-model	4.30	2.64	4.18	2.49	3.73	1.97
Redis	Key-value, Multi-model	11.80	7.39	11.46	6.95	10.23	5.50
Elasticsearch	Search engine, Multi-model	13.70	8.13	13.31	7.65	11.88	6.05
Cassandra	Wide column	10.53	10.75	10.24	10.12	9.14	8.00
Splunk	Search engine	25.75	13.59	25.03	12.79	22.34	10.11
Amazon DynamoDB	Multi-model	47.13	16.51	45.80	15.54	40.88	12.29
Neo4j	Graph	38.41	22.09	37.32	20.78	33.31	16.44
Solr	Search engine, Multi-model	20.50	22.21	19.92	20.89	17.78	16.53
InfluxDB	Time series, Multi-model	241.04	45.16	234.22	42.50	209.04	33.62

The Top-10 relational databases and the Top-5 Key-Value, Document, Wide-Column, Graph and Time-Series are respectively depicted in Tables 1.5-1.10. The tables show the percentage difference in popularity in the last 5 years. It should be mentioned that there are DBMSs that were included in more than one category.

From Tables 1.5-1.10, where the change of the popularity has also been noted, we conclude that from the top-20 Databases, 15 demonstrated an increase in their popularity and the remaining 5, four of which were relational and one a Wide-Column DB, a decrease. The biggest increase in the past 5 years was observed at (a) MariaDB (+136.69%), (b) PostgreSQL (+84.3%) and (c) Splunk (+71.75%) whilst the biggest decrease was observed at: (a) Microsoft SQL Server (−22.23%), (b) MySQL (−12.25%) and (c) Cassandra (−11.23%).

Table 1.5. Top 10 relational DBMS

Rank Dec 2021	DBMS	Additional database model	Score		
			Dec 2021	Dec 2016	Change %
1	Oracle	Multi-model	1,281.74	1,414.405	−9.38
2	MySQL	Multi-model	1,206.04	1,374.406	−12.25
3	Microsoft SQL Server	Multi-model	954.02	1,226.658	−22.23
4	PostgreSQL	Multi-model	608.21	330.016	84.30
5	IBM Db2	Multi-model	167.18	1,84.342	−9.31
6	SQLite		128.68	110.833	16.10
7	Microsoft Access		125.99	124.701	1.03
8	MariaDB	Multi-model	104.36	44.092	136.69
9	Microsoft Azure SQL Database	Multi-model	83.25	20.837	299.53
10	Hive		81.93	49.4	65.85

Table 1.6. Top 5 key-value DBMS

Rank Dec 2021	DBMS	Additional database model	Score		
			Dec 2021	Dec 2016*	Change %
1	Redis	Multi-model	173.54	119.892	44.75
2	Amazon DynamoDB	Multi-model	77.63	30.009	158.69
3	Microsoft Azure Cosmos DB	Multi-model	39.71	3.425	1,059.42
4	Memcached		26.27	28.772	−8.70
5	etcd		10.96	7.148	53.33

* etcd was released in November 2019, therefore the score and the % change is calculated from that month.

Table 1.7. Top 5 document based DBMS

Rank Dec 2021	DBMS	Additional database model	Score		
			Dec 2021	Dec 2016	Change %
1	MongoDB	Multi-model	484.67	328.684	47.46
2	Amazon DynamoDB	Multi-model	77.63	30.009	158.69
3	Microsoft Azure Cosmos DB	Multi-model	39.71	3.425	1059.42
4	Couchbase	Multi-model	28.45	29.728	−4.30
5	Firebase Realtime Database		19.96	2.15	828.37

Table 1.8. Top 5 wide-column DBMS

Rank Dec 2021	DBMS	Additional database model	Score		
			Dec 2021	Dec 2016*	Change %
1	Cassandra		119.2	135.057	−11.74
2	HBase		45.54	58.626	−22.32
3	Microsoft Azure Cosmos DB	Multi-model	39.71	3.425	1,059.42
4	Datastax Enterprise	Multi-model	9.28	5.955	55.84
5	Microsoft Azure Table Storage	Wide column	5.28	2.839	85.98

* Datastax Enterprise was released in February 2018 and Microsoft Azure Table Storage in January 2017, therefore the score and the % change are calculated from those months.

Table 1.9. Top 5 graph DBMS

Rank Dec 2021	DBMS	Additional database model	Score		
			Dec 2021	Dec 2016	Change %
1	Neo4j		58.03	36.827	57.57
2	Microsoft Azure Cosmos DB	Multi-model	39.71	3.425	1,059.42
3	Virtuoso	Multi-model	5.07	2.484	104.11
4	ArangoDB	Multi-model	4.75	2.39	98.74
5	OrientDB	Multi-model	4.4	5.869	−25.03

Table 1.10. Top 5 time-Series DBMS

Rank Dec 2021	DBMS	Additional database model	Score		
			Dec 2021	Dec 2016*	Change %
1	InfluxDB	Multi-model	28.38	5.868	383.64

2	Kdb+	Multi-model	8.1	1.214	567.22
3	Prometheus		6.44	0.348	1,750.57
4	Graphite		5.7	1.899	200.16
5	TimescaleDB	Multi-model	4.38	0.051	8,488.24

* TimescaleDB was released in December 2017, therefore the score and the % change is calculated from that month.

1.3.4 Performance and comparison

M. Sharma et al. [19] compared the performance of search queries done on geolocation data by a Relational DBMS (PostgreSQL), a NoSQL document-based DBMS (MongoDB) and a NoSQL Graph DBMS (Neo4j). They used a volume of 2GB of geospatial data from 600 users. According to the authors they received "unexpected results" in the case of Neo4j as its query execution time was longer compared to MongoDB and PostgreSQL. Two sets of queries were executed, where in the first the number of users (10, 100, 200 and 500 users) and in the second the number of the extracted nodes (100, 500, 1,000 and 5,000 nodes) were increasing gradually. The time taken for Neo4j to perform both sets of queries was dramatically larger compared to PostgreSQL and MongoDB as depicted in Figure 1.6.

In [21] the authors studied Neo4j in the market and evaluated its performance using the Social Network Benchmark (SNB). Their experimental evaluation used two datasets with volumes 1.2 GB and 11.6 GB and it examined the loading time in Neo4j and the SNB Queries execution time. The total loading time for the first dataset was 5 h and 12 minutes and its execution time for 14 queries was 5 minutes, whilst, for the second much larger dataset the loading time was 160.5 h and for the same number of queries approximately 17 minutes were required. Compared to other systems, the authors concluded that Neo4j, "stands out for its simplicity", and even though the users need to have a prior knowledge of Cypher, the Neo4j query language, it is a powerful tool, it is more robust and more practical.

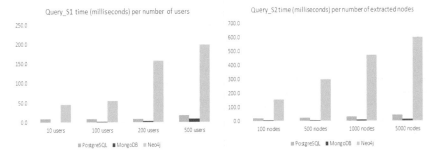

Figure 1.6. Time Query execution comparison (PostgreSQL mongoDB and Neo4j).

Fernandez and Bernandino [23] compared five Graph Databases: AllegroGraph, ArangoDB, InfiniteGraph, Neo4J, and OrientDB, which are listed according to the db-engines.com in December 2021, at the 14th, 4th, 19th, 1st and 5th position at the Graph Databases ranking, respectively. They compared them in 8 different features: (i) Flexible schema, (ii) Query Language, (iii) Sharding, (iv) Backups, (v) Multi-model, (vi) Multi-architecture, (vii) Scalability and (viii) Cloud ready and by grading them with a five Likert scale, from 0 to 4, they concluded that "Neo4J is one of the best options when choosing a graph database and this study shows that this software has the most important features". The grades by features along with the total grade of each of the five Graph Databases are demonstrated at Table 1.11.

Table 1.11. Graph Databases features comparison [23]

	AllegroGraph	ArangoDB	InfiniteGraph	Neo4J	OrientDB
Flexible Schema	1	3	3	4	3
Query Language	3	3	3	4	3
Sharding	3	3	0	0	3
Backups	3	2	3	4	3
Multimodel	4	4	2	2	4
Multi Architecture	3	4	3	4	3
Scalability	3	4	3	4	3
Cloud Ready	3	3	4	4	3
Total	**23**	**26**	**21**	**26**	**25**

Miller [26] investigated the Graph Database Applications like Social Graph, Recommender Systems and Bioinformatics and Concepts with Neo4j and compared it with two RDBMS (Oracle and MySQL). The comparison was based on features like (a) software design, (b) data query and manipulation and (c) reliability and more specifically in terms of transactions, availability, backups and security. Miller concluded that Graph DBs provide "a new way of modeling and traversing interconnected data that is unparalleled in information storage" and they can overcome the limitations that can be found in RDBMS.

An open-source object-graph-mapping Framework for Neo4j and Scala was presented in [34]. It includes two libraries: Renesca, a Graph Database driver that provides graph-query-results and change tracking, and Renesca-magic, a macro-based Entity Relationship modeling Domain Specific Language. Both libraries have been assessed and validated. The results were astounding, as their utilization resulted in a tenfold increase in code size and extensibility, with no notable influence on performance. Holsch and Grossniklaus [37] proposed an optimization of graph algebra and corresponding equivalences to transform graph patterns in Neo4j. Their work focused on "how the relational algebra can be extended with a general Expand operator in order to enable the transformation-based enumeration of different traversal patterns for finding a pattern in a graph". Their contribution is that they managed to demonstrate how their proposed graph algebra can be used to optimize Cypher queries. Holsch and Grossniklaus with the contribution of Schmidt, a year later [38], examined the performance of analytical and pattern matching graph queries in Neo4j and a relational Database. They presented a collection of analytical and pattern matching queries, and they evaluated the performance of Cypher and SQL. Their evaluation showed that the RDBMS outperforms Neo4j for their analytical queries as long as the queries do not need to join the whole edge table multiple times for longer patterns or for queries with cycles, where the performance of SQL was much worse.

1.3.5   Use by the industry

The rise of the popularity of NoSQL DBs is evident by their extensive development and their use from a wide range of colossal enterprises.

MongoDB[7]: Forbes, Toyota, Sanoma, ThermoFisher scientific, KPMG, Vivint., HM revenue & customs, Royal Bank of Scotland, Barclays, Verizon, GAP, Cisco, EA, Adobe, AstraZeneca to name a few.

Redis[8]: Twitter, GitHub, Weibo, Pinterest, Craiglist, Digg, StackOverflow, Flickr and many others.

ElasticSearch[9]: Adobe, T-mobile, Audi, Happyfresh, Zurich, P&G, University of Oxford and many more.

Cassandra[10] Apple, Best Buy, BlackRock, Bloomberg, CERN, Coursera, Ebay, Home Depot, Hulu, Instagram, Macy's, Netflix, the New York Times, Spotify, Walmart amongst many others.

Splunk[11]: Tesco, Zoom, McLaren, The University of Arizona, intel, Honda, Lenovo and many more.

Amazon Dynamo[12]: Amazon, Zoom, Disney, Snap inc., Dropbox, A+E, Netflix, Capital One, Samsung, Nike, tinder, Airbnb amongst others.

Neo4j[13]: COMCAST, UBS, Ebay, LEVI'S, Adobe to name a few.

Finally, solr[14]: AoL, Apple, Cisco, Cnet, Xfinity, Disney, Ebay, Instagram, NASA, Whitehouse and others.

---

[7] https://www.mongodb.com/who-uses-mongodb
[8] https://redis.io/topics/whos-using-redis
[9] https://www.elastic.co/customers/
[10] https://cassandra.apache.org/_/case-studies.html
[11] https://www.splunk.com/en_us/customers.html
[12] https://aws.amazon.com/dynamodb/customers/
[13] https://neo4j.com/who-uses-neo4j/
[14] https://lucidworks.com/post/who-uses-lucenesolr/

1.4   Graph Databases

According to the Wikipedia article about Graph Databases[15] "In computing, a graph database (GDB) is a database that uses graph structures for semantic queries with nodes, edges, and properties to represent and store data. A key concept of the system is the graph (or edge or relationship). The graph relates the data items in the store to a collection of nodes and edges, the edges representing the relationships between the nodes. The relationships allow data in the store to be linked together directly and, in many cases, retrieved with one operation. Graph databases hold the relationships between data as a priority. Querying relationships is fast because they are perpetually stored in the database. Relationships can be intuitively visualized using graph databases, making them useful for heavily inter-connected data".

1.4.1   Popularity of Graph Databases

All 36 databases categorized as Graph Databases according to db-engines or as multi-model, but their primary/main category is Graph Database are cited in Table 1.12. Additional information like their ranking on December 2021, the previous month (November 2021) and the previous year (December 2020) along with the current score and its change (positive, negative or null) are also displayed in Table 1.12.

Table 1.12. Rank and score of Graph Databases [db-engines.com]

Rank			DBMS	Database model	Score		
Dec 2021	Nov 2021	Dec 2020			Dec 2021	Nov 2021	Dec 2020
1	1	1	Neo4j	Graph	58.03	0.05	3.4
2	2	2	Microsoft Azure Cosmos DB	Multi-model	39.71	−1.11	6.17
3	4	6	Virtuoso	Multi-model	5.07	0.25	2.48
4	3	3	ArangoDB	Multi-model	4.75	−0.35	−0.76
5	5	4	OrientDB	Multi-model	4.4	−0.24	−0.89
6	6	8	GraphDB	Multi-model	2.88	0.05	0.84
7	7	7	Amazon Neptune	Multi-model	2.56	−0.04	0.05
8	8	5	JanusGraph	Graph	2.41	−0.13	−0.24
9	9	12	TigerGraph	Graph	2	−0.03	0.53
10	10	11	Stardog	Multi-model	1.93	−0.04	0.47
11	11	10	Dgraph	Graph	1.58	−0.11	−0.1

(Conted.)

[15] https://en.wikipedia.org/wiki/Graph_database

Table 1.12. (*Conted.*)

12	12	9	Fauna	Multi-model	1.4	−0.23	−0.62
13	13	13	Giraph	Graph	1.35	−0.01	0.21
14	14	14	AllegroGraph	Multi-model	1.23	0	0.16
15	15	15	Nebula Graph	Graph	1.14	0	0.28
16	16	16	Blazegraph	Multi-model	0.96	0.03	0.12
17	17	18	Graph Engine	Multi-model	0.85	0.01	0.15
18	18	17	TypeDB	Multi-model	0.78	−0.02	0.02
19	19	19	InfiniteGraph	Graph	0.46	−0.02	−0.04
20	20	28	Memgraph	Graph	0.37	0.01	0.29
21	21	26	AnzoGraph DB	Multi-model	0.34	0.02	0.2
22	22	21	FlockDB	Graph	0.3	0	−0.03
23	23	20	Fluree	Graph	0.24	−0.01	−0.12
24	24	27	TerminusDB	Graph, Multi-model	0.24	0	0.13
25	25	22	HyperGraphDB	Graph	0.2	−0.02	−0.07
26	26	25	TinkerGraph	Graph	0.15	−0.01	−0.02
27	27	23	Sparksee	Graph	0.12	0	−0.06
28	29	30	HugeGraph	Graph	0.12	0.03	0.05
29	28		ArcadeDB	Multi-model	0.11	0	
30	30	24	GraphBase	Graph	0.08	0.01	−0.1
31	31	31	AgensGraph	Multi-model	0.07	0.01	0
32	32	29	VelocityDB	Multi-model	0.04	0	−0.04
33	33	33	Bangdb	Multi-model	0.04	0	0.04
34	34		RDFox	Multi-model	0.03	0	
35	35	32	HGraphDB	Graph	0.02	0	0.01
36	36		Galaxybase	Graph	0.02	0.02	

According to the db-engines.com ranking (Table 1.12) the most popular Graph DBMS are Neo4j (#19 ranking amongst all the DBMSs) and Microsoft Azure Cosmos DB (#26 ranking amongst all the DBMSs). The next three most popular Graph DBMSs, Virtuoso, ArangoDB and OrientDB ranked at the 72nd, 76th and 80th respectively, have quite low scores. From the above statistics the highest increase in popularity, in the past year, is observed by Microsoft Azure Cosmos DB, followed by Neo4j and then by Virtuoso.

Figure 1.7 shows the ranking trends of the top-15 Graph DBMS as provided by db-engines.com. It is obvious that Neo4j and MS Azure Cosmos DB are the dominant Graph DBMS and their popularity is on a constant increase.

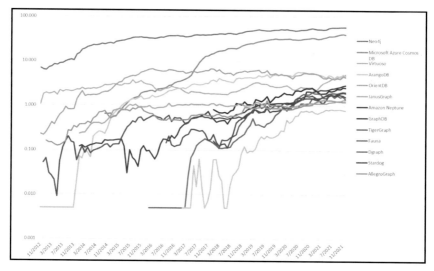

Figure 1.7. DB-engines ranking of Graph DBMS.

1.5   Neo4j

According to the www.neo4j.com, Neo4j is "The Fastest Path to Graph and it gives developers and data scientists the most trusted and advanced tools to quickly build today's intelligent applications and machine learning workflows. Available as a fully managed cloud service or self-hosted". As mentioned in db-engines.com Neo4j System Properties[16], Neo4j supports a range of programming languages like .Net, Clojure, Elixir, Go, Groovy, Haskell, Java, JavaScript, Perl, PHP, Python, Ruby and Scala.

Neo4j's key competitive advantages are:

- 1000x Performance at Unlimited Scale
- Unmatched Hardware Efficiency
- Platform Agnostic
- Developer Productivity - Declarative Language
- Agility - Flexible Schema
- First Graph ML for Enterprise

From Figure 1.8 and Table 1.13 that depict the ranking of Neo4j from November 2012 until December 2021, it is evident that Neo4j is in a constant rise, demonstrating a big 614% increase between December 2012 and December 2021. It is also worth mentioning that each year Neo4j demonstrates an increase on its popularity compared to the previous year. The last six years the increase varies from 5.16% (December 2017) up to 17.63% (December 2018).

---

[16] https://db-engines.com/en/system/Neo4j

Figure 1.8. Neo4j ranking (2012-2021) (db-engines.com).

Table 1.13. Neo4j change in ranking (Dec. 2012-Dec. 2021)

December	Score	Change (previous year) %	Change (December 2012) %
2012	6.38	-	-
2013	13.773	115.88	115.88
2014	25.168	82.73	294.48
2015	33.183	31.85	420.11
2016	36.827	10.98	477.23
2017	38.727	5.16	507.01
2018	45.556	17.63	614.04
2019	50.56	10.98	692.48
2020	54.636	8.06	756.36
2021	58.034	6.22	809.62

1.5.1   Neo4j Customers

As stated by Neo4j's website, 7 of the World's Top 10 Retailers like ADEO, eBay, and ATPCO, 3 of the Top 5 Aircraft Manufacturers like Airbus, 8 of the Top 10 Insurance Companies like die Bayerische and Allianz, all of North America's Top 20 Banks like JPMorgan Chase, Citi, and UBS, 8 of the Top 10 Automakers like Volvo Cars, Daimler and Toyota, 3 of the World's Top 5 Hotels like Marriott and AccorHotels, and 7 of the Top 10 Telcos like Verizon, Orange, Comcast, and AT&T rely on Neo4j:

- to drive recommendations, promotions, and streamline logistics,
- to connect and solve complex data problems,

- to manage information and fight fraud,
- for data lineage, customer 360 and regulatory compliance,
- to drive innovative manufacturing solutions,
- to manage complex and rapidly changing inventories, and
- to manage networks, control access, and enable customer 360.

According to db-engines.com there are over 800 commercial customers and over 4,300 startups using Neo4j. To complement the above list of customers, the website mentions that flagship customers also include Walmart, Cisco, Citibank, ING, HP, CenturyLink, Telenor, TomTom, Telia, Scripps Interactive Networks, The National Geographic Society, DHS, US Army, Pitney Bowes, Vanguard, Microsoft, IBM, Thomson Reuters, Amadeus Travel, Caterpillar, and many more.

In addition, Neo4j "boasts the world's largest graph database ecosystem with more than 140 million downloads and Docker pulls, over 200K active instances, and 800+ enterprise customers, including over 75% of Fortune 100 companies".

The reason behind this trust is that "Neo4j is the clear performance leader in graph technology", many data and Graph scientists and developers are using Neo4j as its tools and functionalities along with its easy-to-use graph analytics are boosting the production reliability, flexibility, and integrity for high-volume transaction/analytic workloads.

1.5.2 Neo4j Use Cases and Scientific Research

According to db-engines.com[17] the typical application scenarios for Neo4j are:

- Real-Time Recommendations
- Master Data Management
- Identity and Access Management
- Network and IT Operations
- Fraud Detection
- Anti-Money Laundering/Tax Evasion
- Graph-Based Search
- Knowledge Graphs
- Graph Analytics and Algorithms
- Graph-powered Artificial Intelligence
- Smart Homes
- IoT/Internet of Connected Things

On its webpage about the use cases[18] the developers state that "early graph innovators have already pioneered the most popular use cases" and that "building upon their efforts, the next generation of graph thinkers are engineering the future of artificial intelligence and machine learning".

---

[17] https://db-engines.com/en/system/Neo4j
[18] https://neo4j.com/use-cases/

The most popular use cases, as defined by the Neo4j developers, are:

1. Fraud detection and analytics: Real-time analysis of data connections is critical for identifying fraud rings and other sophisticated schemes before the fraudsters and criminals do long-term damage.
2. Network and database infrastructure monitoring for IT operations: Graph databases are fundamentally better suited than RDBMS for making sense of the complex interdependencies that are key to network and IT infrastructure management.
3. Recommendation engine and product recommendation system: Graph-powered recommendation engines assist businesses in personalizing products, information and services by utilizing a large number of real-time connections.
4. Master data management: Organize and manage the master data using a flexible and schema-free graph database format in gain to get real-time insights and a 360° view of the customers.
5. Social media and social network graphs: When employing a graph database to power a social network application, you may easily utilize social connections or infer associations based on activity.
6. Identity and access management: When utilizing a graph database for identity and access management, you may quickly and efficiently track people, assets, relationships, and authorizations.
7. Retail: For retailers, Neo4j supports real-time product recommendation engines, customer experience personalisation, and supply-chain management.
8. Telecommunications: Neo4j assists the telecom sector in managing complex interdependencies in telecommunications, IT infrastructure, and other dense networks.
9. Government: Governments utilize Neo4j to combat crime, prevent terrorism, enhance fiscal responsibility, increase efficiency, and provide transparency to their citizens.
10. Data privacy, risk and compliance: Regulatory compliance that is quick and effective (GDPR, CCPA, BCBS 239, FRTB and more). Neo4j assists in the management of enterprise risk while harnessing linked data to improve business intelligence.
11. Artificial intelligence and analytics: Artificial Intelligence is expected to lead the next wave of technology disruption in practically every sector.
12. Life sciences: Pharmaceutical, chemical, and biotech businesses are adopting Neo4j to examine data in ways that were previously impossible without graphics.
13. Financial services: Context is essential in everything from risk management to securities advice. Top banks throughout the world are utilizing Neo4j to address their linked data concerns.
14. Graph data science: Businesses now confront tremendously complex issues and possibilities that necessitate more adaptable, intelligent ways. That's

why Neo4j built the first enterprise graph framework for data scientists: to enhance forecasts, which lead to better decisions and innovation.

15. Supply chain management: Graph technology is critical for optimizing products movement, identifying weaknesses, and increasing overall supply chain resilience. Explore how Transparency-One, Caterpillar, and other companies are using supply chain graph technology to secure business continuity.

16. Knowledge graph: Knowledge graphs guarantee that search results are contextually appropriate to the needs of the user. Knowledge graphs are used by organizations like NASA, AstraZeneca, NBC News, and Lyft to contextualize a range of data forms and formats.

1.5.3   Neo4j Research

There are many academic, scientific, research articles that study the use of Neo4j. In this section, a literature review is going to be performed and the most important findings are going to be mentioned. Each article is going to be categorized in one, or more, if applicable, of the above 16 Neo4j use cases.

Network and Database Infrastructure Monitoring for IT Operations (#2)

Summer et al. [30] created the cyNeo4j application to connect Cytoscape, an open-source software platform for visualizing complex networks and combining them with any sort of attribute data but also with Neo4j. They utilized a subset of the STRING network[19] with 4,436 nodes and 93,286 edges, and they were able to improve the efficiency of network analysis methods and decrease calculation time by a factor of four when compared to the Cytoscape version.

Drakopoulos suggested a tensor Fusion of Social Structural and Functional Analytics over Neo4j which was applied to a Twitter subgraph with the hashtag #brexit [35]. The author showcased harmonic centrality, a structural metric for ranking vertices of large graphs and provided metrics (i) on basic information structural and functional information about the dataset used in the research, (ii) #brexit related hashtags, (iii) Zipf exponent estimation, that is based on Zipf's law which states that the frequencies of certain events are inversely proportional to their rank, and (iv) data quality metric values. The research can alternatively be labeled under the "Social media and social network graphs" category.

Recommendation Engine and Product Recommendation System (#3)

Virk and Rani on their paper titled "Efficient approach for social recommendations using graphs on Neo4j" [17] are proposing a recommendation system based on Neo4j and a transitivity in social media that enables the encoding of larger volumes of data. They have used the FilmTrust dataset and according to their results, they outperformed state of the art recommendation systems, they overcame problems of traditional recommendation systems, they solved sparsity problems and they

---

[19] https://string-db.org/

managed to increase the trust between users. This research can also be classified as a "Social media and social network graphs" use case scenario.

Bakkal et al. [27] suggested a unique carpooling matching approach. The authors modeled trajectories using the Neo4j spatial and TimeTree libraries, then performed temporal and locational filtering stages, and ultimately evaluated the efficiency and efficacy of the proposed system. The suggested approach showed available drivers to hitchhikers using Google Maps and presented them in the form of a graph using a Geolife trajectory dataset of 182 individuals and 17,621 trajectories. According to the assessment, the system was effective, and it could assist carpooling by saving users money on gasoline, tolls, and time wasted on the road, among other things.

Another research attempt [28] proposes a recommendation system for impacted items based on association rule mining and Neo4j. Using real-world consumer feedback data from Amazon, the authors created a model to determine the effect of one product on another for better and faster decision making. According to the findings, the model outperforms the Apriori algorithm, one of the most common approaches for association rule mining. It has lower time complexity and saves space, whereas the Apriori methodology has a much higher runtime complexity than the Neo4j model. This paper also falls under the category of a "Retail" use case scenario.

Dharmawan and Samo [33] proposed a book recommendation system by combining BibTeX book metadata and Neo4j. Then, with the aid of Cypher by inputting criteria like the author's name or the book's type, the user can receive book recommendation based on their input's criteria. They performed various queries with both SQL and Cypher and the results were the same. However, Neo4j queries took around 130 milliseconds on average to be executed for the author's criteria and 124 milliseconds for the book's type criteria, but they were approximately 8.5 and 10 times slower, respectively, than the same queries executed in SQL. Despite this difference in execution time, that does not actually affect the user as they are at the range of hundredth of a second, the authors concluded that storing book recommendation into database, graph database is more flexible than a relational database and graph databases are more efficient in preparing data before querying them and more flexible for a book recommendation system. This paper can also be classified as a "Retail" use case.

Konno et al. [41] proposed a goods recommendation system based on retail knowledge in a Neo4j Graph Database combined with an inference mechanism implemented in Java Expert System Shell (Jess). They presented a two-layer knowledge graph database, a concept layer, "the resulting graph representation transferred from an ontology representation" and an instance layer, "the instance data associated with concept nodes". By using RFM (Recency, Frequency, Monetary) analysis, "they extracted customer behavior characteristics and classified customers into five groups according to RFM segments and a list of recommended products was created for each segment based the RFM value associated to each customer". Finally, they evaluated the time efficiency of answering queries of retail data and the novelty of recommendations of the system and they concluded that they were "reasonably good". This study can also be classified as a "Retail" use case.

Finally, the authors in [44] exploited Neo4j to create a content-based filtering recommendation system in abstract search. They designed a recommendation system based on the search results for related documents based on the content search on report papers. They employed Neo4j, and a document-keyword graph was developed to depict the relationship between the document and its characteristics, and it was used to filter keyword co-occurrence documents in order to limit the search space as much as feasible. The model's efficiency was examined, and the findings showed that it had an accuracy of 0.77.

Master Data Management (#4)

In [31] the authors developed a Model Driven Approach for reverse engineering of NoSQL property graph databases using Neo4j. They demonstrated an exemplary situation and assessed their four-step reverse engineering technique. The four steps were: (i) collecting all Cypher codes that enabled the Neo4j graph database generation; (ii) parsing the code to determine the logical graph using transformation rules; (iii) mapping the logical graph into a conceptual Extended Entity-Relationship schema using transformation rules; and (iv) completing the resulting schema by adding the relationships. Their approach provided the Neo4j database user with an automatic generation of the property graph as well as an automatic generation of a conceptual representation, which the authors claim that can be easily extended and applied for reverse engineering of other NoSQL DBMS, such as key-value or document based.

Social Media and Social Network Graphs (#5)

Constantinov et al. [20] performed an analysis with Neo4j on social data related on how University graduates were integrated in the labor market and therefore investigated workforce trends and corporate information leakage. With the aid of Neo4j they retrieved and processed information about (i) job type distribution, (ii) showcased skills, (iii) endorsed skills, (iv) companies by the number of active jobs (employees), (v) top endorsed skills for the selected companies, (vi) average time (in months) an employee spends with the company and (vii) the leaves from one company to another. Their proposed system, which provided various data modelling techniques along with visualization techniques on professional SMN enabled the uncovering of valuable feedback and insights from within the above data. Finally, the authors highlighted how a group of employees might unwittingly disclose company information on social media.

Community discovery or detection is "a fundamental problem in network science which aims to find the most likely community that contains the query node"[20] and it has attracted great attention from both academic and industry areas.

---

[20] https://en.wikipedia.org/wiki/Community_search

In [22] the authors presented the implementation of four established community discovery algorithms with the aid of Neo4j and a local scale-free graph behavior was proposed and tested on the results of applying these algorithms to two real Twitter graphs created from a neutral as well as a politically charged topic.

Cattuto et al. [24] introduced a data model for data retrieved from time-varying social networks that were represented as a property graph in the Neo4j DBMS. They collected data by using wearable sensors and they studied the performance of real-world queries, pointing to strengths, weaknesses, and challenges of the proposed approach. They observed that the combination of their chosen data models and Neo4j performs very well when querying the data by performing exploratory data analysis, and research-oriented data mining. They also suggested that in data from social media structures is better to connect nodes by using "strongly-typed" relationships.

In their paper, Allen et al. [29] tried to understand social media "Trolls", persons who make a deliberately offensive or provocative online posts, with Efficient Analytics of Large Graphs in Neo4j. The authors described the design and the integration of Neo4j Graph Algorithms, and they demonstrated its utility of their approach with a Twitter Troll analysis and showcased its performance with a few experiments on large graphs. They ran four algorithms, PageRank, Union Find, Label Propagation, and Strongly Connected Components (SCC) on seven graph datasets: Pokec, cit-patents, Graphs500-23, soc-LifeJournal1, DBPedia, Twitter-2010 and Friendster that totaled 118.42 GB of data, contained 133.59 million nodes and 3,636.5 million relationships. They measured the total runtime of the complete procedure call, which included loading the graph, computing the analysis, and writing back the result. Their work can aid the uncovering of relationships of interest that are hidden behind layers of many other connections and demonstrated the power of graph analytical algorithms, of Cypher and Neo4j in general.

In [36] the authors defined and evaluated Twitter influence metric with the aid of Neo4j. They examined seven first-order influence metrics for Twitter, defined a strategy for deriving their higher-order counterparts, and outlined a probabilistic evaluation framework. Their framework that was implemented in the Python ecosystem with probabilistic tools from various fields including information theory, data mining, and psychometrics and it can be used for assessing the performance of an influence metric in various social media networks.

Retail (#7)

In [42] a study about the Neo4j efficient storage performance of oilfield ontology was conducted. The authors designed mapping rules from ontology files to regulate the Neo4j database and via a two-tier index architecture, that included object and triad indexing, they managed to initially to keep the time of loading data low and they were able to match with different patterns for accurate retrieval.

Their evaluation results were very promising, and they showed that Neo4j can reduce the required storage space in a great extent, since their method could "save 13.04% of the storage space and improve retrieval efficiency by more than 30 times compared with the methods of relational databases".

Life Sciences (#12)

In [18] the authors pointed out that the traditional relational databases, when using a very large number of links between data, are making the querying of the database cumbersome and troublesome. Therefore, they implemented Neo4j in order to analyze, query and visualize data from various diseases. Their goal was to provide a better and more intuitive image analysis, compared to traditional relational databases and to suggest corresponding treatments for the disease data of related queries. They concluded that Neo4j is more suitable for large, complex and low structured data and that the retrieved results from the queries provided information that can help the patients to find the more suitable treatment.

Stothers and Nguyen [25] wondered if Neo4j can replace PostgreSQL in Healthcare. They compared the query languages of the two DBMSs, Cypher and SQL respectively, in terms of Clause Complexity (the average number of clauses per operation) and Operation Runtime (the average operation time). They found that regarding the same commands, Cypher needed 30% fewer clauses (7.86) compared to SQL (10.33). In addition, Neo4j was substantially faster than PostgreSQL as the average operation time of Cypher was 45.25 ms whilst SQL's was around 3.5 times slower, averaging 154.48 ms. Based on the previous results the authors concluded that, Neo4j queries are less complex and have a faster runtime than PostgreSQL and despite the fact that Neo4j is a bit more difficult to implement, "Neo4j should be considered a viable contender for health data storage and analysis".

The study by Soni et al. [32] focused on a use case of graph DBMS in healthcare by utilizing Neo4j to store and analyze tweets containing at least one opioid-related term. Their investigation provided information to healthcare and government officials about the public usage of opioids and the geographical distribution of opioid-related tweets. Some of their findings were: (i) that California, and particularly the Sacramento area, had the highest number of users (921 users) who posted 2,397 opioid-related tweets during the period of the research, (ii) that North Carolina had the highest percentage (17%) of opioid-related tweets when compared to the total number of the state's tweets, (iii) that the largest opioid-related user group on Twitter had 42 users and (iv) the most discussed topic in the group was the adverse effects of Tylenol and Percocet. The findings could help healthcare and government professionals by giving important information to tailor public health campaigns to populations in high-use regions and to notify authorities about locations with significant opioid distribution and usage. This document could also be characterized as a "Government" use case scenario.

In [39] the authors used Neo4j for mining protein graphs adding another study about the usefulness of applying Neo4j in bioinformatics and in health sciences in general. The problem they managed to resolve was the one of protein-protein interface (PPI) identification in which according to the authors "the goal of the PPI identification task is, given a protein structure, to identify amino acids which are responsible for binding of the structure to other proteins". They retrieved data from the Protein Data Bank (PDB) and showed a method for transforming and migrating these data into Neo4j as a set of independent protein graphs. The resulting graph database contains about 14 million labeled nodes and 38 million edges and after querying the graph database they concluded that "using Neo4j is a viable option for specific, rather small, subgraph query types" but at the same time querying large number of edges led to performance limitation.

Another study on proteins was undertaken by Johnpaul and Mathew [40], who presented a Cypher Query based NoSQL Data Mining on Protein Datasets using Neo4j Graph Database. The research investigated the usage of Cypher queries on a proteome-protein dataset of 20,000 nodes and 100,000 edges, with an average degree of five for each node. The assessment was based on inquiries with varying total number of nodes and relationships between them. The research concluded by claiming that NoSQL queries are more capable for conducting data retrieval without the restrictions of RDMSs and they offer better storing for unstructured data.

Stark et al. [43] presented a migraine drug recommendation system based on Neo4j, called BetterChoice. Their proposed system used simulated data for 100,000 patients to help physicians gain more transparency about which drug fits a migraine patient best considering his/her individual features. They evaluated the system by examining if the recommended drugs best suited the patients' needs and they concluded that "the proposed system works as intended as only drugs with highest relevance scores and no interactions with the patient's diseases, drugs or pregnancy were recommended".

1.6 Discussion

This chapter compares the RDs with the NoSQL DBs, cites their main advantages and disadvantages and investigates their popularity, based on a methodology introduced by the db-engines website. In this platform 432 database management systems are analyzed, and 381 of them are distributed in 15 categories and ranked based on their popularity. After statistical analysis, it was concluded that the Relational Databases continue to dominate the DBMS domain, as approximately 40% of the analyzed DBMS belong to this category. The categories that follow are Key-Value stores with 16.8%, Document stores with 13.91%, Time Series DBMS with 10.24% and Graph DBMS with 9.45%. The remaining 10 categories claim less than 6% of the total number of analyzed DBMS.

However, the gap in popularity between RDBMS and the non-RDBMS has been reducing as the top-3 RDBMS are becoming less popular, whereas the non-

Table 1.14. Distribution of papers per use-case category

Reference no.	Publi-cation year	Comparison - Performance	Net-works (#2)	Recom-mendation (#3)	Data manage-ment (#4)	Social media (#5)	Retail (#7)	Govern-ment (#9)	Life sciences (#12)
[17]	2018			1		1			
[18]	2019								1
[19]	2018	1							
[20]	2018	1				1			
[21]	2018	1				1			
[22]	2017	1				1			
[23]	2018	1							
[24]	2013					1			1
[25]	2020								
[26]	2013	1							
[27]	2017			1					
[28]	2021			1			1		
[29]	2019					1			
[30]	2015		1						
[31]	2017				1				
[32]	2019							1	
[33]	2017			1		1	1		1

	Year	25.00%	7.14%	25.00%	3.57%	25.00%	14.29%	3.57%	21.43%
[34]	2016	1							
[35]	2016		1			1			
[36]	2017					1			
[37]	2016	1							
[38]	2017	1							
[39]	2015								1
[40]	2017								1
[41]	2017			1			1		
[42]	2018						1		
[43]	2017			1					1
[44]	2017			1					
Total 28		7	2	7	1	7	4	1	6
Percentage		**25.00%**	**7.14%**	**25.00%**	**3.57%**	**25.00%**	**14.29%**	**3.57%**	**21.43%**

Table 1.15. Total and average number of citations per paper

Reference no.	Publication year	Citations	Citations per year	Reference no.	Publication year	Citations	Citations per year
[17]	2018	6	1.5	[31]	2017	26	5.2
[18]	2019	2	0.67	[32]	2019	1	0.33
[19]	2018	16	4	[33]	2017	13	2.6
[20]	2018	5	1.25	[34]	2016	15	2.5
[21]	2018	27	6.75	[35]	2016	19	3.17
[22]	2017	15	3	[36]	2017	25	5
[23]	2018	62	15.5	[37]	2016	26	4.33
[24]	2013	78	8.67	[38]	2017	20	4
[25]	2020	4	2	[39]	2015	10	1.43
[26]	2013	350	38.89	[40]	2017	7	1.4
[27]	2017	8	1.6	[41]	2017	8	1.6
[28]	2021	2	2	[42]	2018	21	5.25
[29]	2019	7	2.33	[43]	2017	9	1.8
[30]	2015	18	2.57	[44]	2017	9	1.8

RDBMS are becoming more popular every year. Then, graph databases and more specifically Neo4j were presented. By examining the literature concerning Neo4j papers that were retrieved from Google Scholar by searching for the term "Neo4j use cases", without applying a date filter and by selecting them based on the order of the search results, it was evident that Neo4j is a quite popular to research and study software. At Table 1.14, the distribution of papers per use-case category (introduced in section "Neo4j Use Cases and scientific research") is demonstrated.

Out of the 28 papers, 7 (25%) focused on: (i) evaluating Neo4j's features and/ or comparing it with other DBMS, (ii) recommendation systems, and (iii) utilizing it with social media, followed by 6 (21.4%) which focused on life sciences use-cases. Since several papers were included in more than one use-case category, summing up percentages exceeds 100%.

Neo4j was released in 2007, however research on Neo4j is recent, as shown by their publication year. The total number of papers since 2017 were 21 (75%). More specifically, 10 were published in 2017, 6 in 2018, 3 in 2019 and 1 in 2020 as well as 2021. No papers published before 2013 were analyzed.

In Table 1.15, the total number of citations along the average number of citations per year, for each article is depicted.

The total number of citations (cited by other papers according to Google Scholar) for the 28 papers was 809, although 350 (43.26%) of them, belonged to a single paper [26]. By finding the average number of citations per year, dividing each paper's citations by the years since the paper was published (2021 − publication year +1), it was calculated that the 28 papers average 131.14 citations per year.

Table 1.16. Average citations per year, and per year per paper

Average	Comparison - Performance	Recommendation (#3)	Social media (#5)	Retail (#7)	Life sciences (#12)
Per Year	76.0	12.9	24.9	11.5	7.6
Per Year - Per Paper	10.9	1.8	3.6	2.9	1.3

The most popular ones are [26] ("Graph database applications and concepts with Neo4j"), by Miller with 350 citations and approximately 39 per year, followed by [24] ("Time-varying social networks in a graph database: a Neo4j use case"), by Cattuto et al. with 78 citations and around 9 per year, followed by [23] ("Graph Databases Comparison: AllegroGraph, ArangoDB, InfiniteGraph, Neo4J, and OrientDB"), by Fernandes & Bernardino, with 62 citations, with 15.5 average per year. It should be noted that 2 of the 3 papers focused on comparing the performance of Graph Databases.

Finally, the most popular use-cases were also analyzed. The threshold was four, meaning that only the categories with four or more citations were examined, and these were: (i) comparison, (ii) recommendation systems, (iii) social media, (iv) retail, and (v) life sciences. The total average citations per year and the average per year and paper are shown in Table 1.16.

This chapter can be used in a bibliographic survey as information and statistics about 381 DBMS, along with the most popular ones per category, and more specifically about relational, key-value, wide-column, document-based, graph and time series DBMS are presented and discussed.

Furthermore, it can be used to promote the use of NoSQL, Graph Databases and mainly Neo4j. It compares Relational Databases and NoSQL Databases and focuses mainly on the numerous advantages of the NoSQL Databases. Their characteristics and their categorization were also studied. Several studies that were analyzed echo the belief that NoSQL Databases, and mainly Graph Databases, outperform the Relational Databases. This conclusion is also backed up by the fact that NoSQL Databases demonstrate a continuous increase on their popularity, as studied and calculated by db-engines.com, and at the same time by the trust of an excessive number of colossi in numerous industries. The db-engines platform's methodology, its categorization of the 381 DBMSs, its measuring popularity, the RDBMS vs. NoSQL popularity, the most popular Relational, Key-Value, Document, Wide-Column, Graph and Time-Series databases were presented.

Studies about the comparison and the performance of RD and NoSQL databases and the use of NoSQL databases were also discussed. Then a more extensive study of Graph Databases, and mainly on the leading one, Neo4j was performed. This study of Neo4j also can serve as a literature review since it offers the analysis of 28 papers related to Neo4j use cases, retrieved from Google Scholar, along with data about the category (out of 16 categories that were introduced on the study) of use-case they investigated and statistics about their popularity by the research community.

Because of the daily engagement of the authors with Neo4j, this chapter could be considered as a proselytism to Neo4j and its awards in 2021 and 2022 Data Breakthrough Award Winners, KMWorld 100 Companies That Matter in Knowledge Management 2020-2022, DSA Awards 2021, The insideBIGDATA IMPACT 50 List for Q4 2020, Q1-Q4 2021 and Q1 2022, 2021 Datanami Readers' Choice Awards, the 2021 SaaS Awards, The Coolest Database System Companies Of The 2021 Big Data 100, Management 2021, Digital.com Names Best Database Management Software of 2021, Neo4j is a Finalist in 2020-21 Cloud Awards, Best Graph Database: 2020 DBTA Readers' Choice Award, KMWorld AI 50: The Companies Empowering Intelligent Knowledge Management, 13th Annual Ventana Digital Innovation Award Finalist: Data, The Coolest Database System Companies Of The 2020 Big Data 100, Neo4j Wins 2020 Data Breakthrough Award for Overall Open Source Data Solution Provider of the Year, Neo4j in The 2019 SD Times 100: DATABASE AND DATABASE MANAGEMENT, Bloor Mutable Award 2019, Neo4j: 2019 Big Data 100 can back it up.

References

[1]   Codd, E.F. 1972. Relational Completeness of Data Base Sublanguages (pp. 65-98). IBM Corporation.

[2]   Seed Scientific. 2021, October 28. How Much Data Is Created Every Day? [27 Powerful Stats]. Retrieved November 1, 2021, from https://seedscientific.com/how-much-data-is-created-every-day/.

[3]   21 Big Data Statistics & Predictions on the Future of Big Data, June 2018 [online]. Available: https://www.newgenapps.com/blog/big-data-statisticspredictions-on-the-future-of-big-data.

[4]   NoSQL. 2020, August 1. Retrieved August 4, 2021, from https://en.wikipedia.org/wiki/NoSQL.

[5]   Moniruzzaman, A.B.M. and S.A. Hossain. 2013. Nosql database: New era of databases for big data analytics-classification, characteristics and comparison. arXiv preprint arXiv:1307.0191.

[6]   Rousidis, D., P. Koukaras and C. Tjortjis. 2020, December. Examination of NoSQL transition and data mining capabilities. *In:* Research Conference on Metadata and Semantics Research (pp. 110–115). Springer, Cham.

[7]   Vaghani, R. 2018, December 17. Use of NoSQL in Industry. Retrieved August 5, 2021, from https://www.geeksforgeeks.org/use-of-nosql-in-industry.

[8]   Nayak, A., A. Poriya and D. Poojary. 2013. Type of NOSQL databases and its comparison with relational databases. *International Journal of Applied Information Systems*, 5(4): 16–19.

[9]   NoSQL Databases List by Hosting Data – Updated 2020. (2020, July 03). Retrieved August 5, 2021, from https://hostingdata.co.uk/nosql-database/.

[10] Zollmann, J. 2012. Nosql databases. Retrieved from Software Engineering Research Group: http://www. webcitation. org/6hA9zoqRd.

[11] DeCandia, G., D. Hastorun, M. Jampani, G. Kakulapati, A. Lakshman, A. Pilchin and W. Vogels. 2007. Dynamo: Amazon's highly available key-value store. *ACM SIGOPS Operating Systems Review*, 41(6): 205–220.

[12] Chang, F., J. Dean, S. Ghemawat, W.C. Hsieh, D.A. Wallach, M. Burrows and R.E. Gruber. 2008. Bigtable: A distributed storage system for structured data. *ACM Transactions on Computer Systems (TOCS)*, 26(2): 1–26.

[13] Koukaras, P., C. Tjortjis and D. Rousidis. 2020. Social media types: Introducing a data driven taxonomy. *Computing*, 102(1): 295–340.

[14] Koukaras, P. and C. Tjortjis. 2019. Social media analytics, types and methodology. *In:* Machine Learning Paradigms (pp. 401-427). Springer, Cham.

[15] Rousidis, D., P. Koukaras and C. Tjortjis. 2020. Social media prediction: A literature review. *Multimedia Tools and Applications*, 79(9): 6279–6311.

[16] Koukaras, P., D. Rousidis and C. Tjortjis. 2020. Forecasting and prevention mechanisms using social media in health care. *In:* Advanced Computational Intelligence in Healthcare-7 (pp. 121–137). Springer, Berlin, Heidelberg.

[17] Virk, A. and R. Rani. 2018, July. Efficient approach for social recommendations using graphs on Neo4j. *In:* 2018 International Conference on Inventive Research in Computing Applications (ICIRCA) (pp. 133–138). IEEE.

[18] Zhao, J., Z. Hong and M. Shi. 2019, June. Analysis of disease data based on Neo4J graph database. *In:* 2019 IEEE/ACIS 18th International Conference on Computer and Information Science (ICIS) (pp. 381–384). IEEE.

[19] Sharma, M., V.D. Sharma and M.M. Bundele. 2018, November. Performance analysis of RDBMS and no SQL databases: PostgreSQL, MongoDB and Neo4j. *In:* 2018 3rd International Conference and Workshops on Recent Advances and Innovations in Engineering (ICRAIE) (pp. 1–5). IEEE.

[20] Constantinov, C., L. Iordache, A. Georgescu, P.Ş. Popescu and M. Mocanu. 2018, October. Performing social data analysis with Neo4j: Workforce trends & corporate information leakage. *In:* 2018 22nd International Conference on System Theory, Control and Computing (ICSTCC) (pp. 403–406). IEEE.

[21] Guia, J., V.G. Soares and J. Bernardino. 2017, January. Graph Databases: Neo4j analysis. *In:* ICEIS (1) (pp. 351–356).

[22] Kanavos, A., G. Drakopoulos and A.K. Tsakalidis. 2017, April. Graph community discovery algorithms in Neo4j with a regularization-based evaluation metric. *In:* WEBIST (pp. 403–410).

[23] Fernandes, D. and J. Bernardino. 2018, July. Graph databases comparison: AllegroGraph, ArangoDB, InfiniteGraph, Neo4J, and OrientDB. *In:* Data (pp. 373–380).

[24] Cattuto, C., M. Quaggiotto, A. Panisson and A. Averbuch. 2013, June. Time-varying social networks in a graph database: A Neo4j use case. *In:* First International Workshop on Graph Data Management Experiences and Systems (pp. 1–6).

[25] Stothers, J.A. and A. Nguyen. 2020. Can Neo4j replace PostgreSQL in healthcare? AMIA Summits on Translational Science Proceedings, 2020, 646.

[26] Miller, J.J. 2013, March. Graph database applications and concepts with Neo4j. *In:* Proceedings of the Southern Association for Information Systems Conference, Atlanta, GA, USA (Vol. 2324, No. 36).

[27] Bakkal, F., S. Eken, N.S. Savaş and A. Sayar. 2017, July. Modeling and querying trajectories using Neo4j spatial and TimeTree for carpool matching. *In:* 2017 IEEE International Conference on INnovations in Intelligent SysTems and Applications (INISTA) (pp. 219–222). IEEE.

[28] Sen, S., A. Mehta, R. Ganguli and S. Sen. 2021. Recommendation of influenced products using association rule mining: Neo4j as a case study. *SN Computer Science*, 2(2): 1–17.

[29] Allen, D., A. Hodler, M. Hunger, M. Knobloch, W. Lyon, M. Needham and H. Voigt. 2019. Understanding trolls with efficient analytics of large graphs in Neo4j. *BTW* 2019.

[30] Summer, G., T. Kelder, K. Ono, M. Radonjic, S. Heymans and B. Demchak. 2015. cyNeo4j: Connecting Neo4j and Cytoscape. *Bioinformatics*, 31(23): 3868–3869.

[31] Comyn-Wattiau, I. and J. Akoka. 2017, December. Model driven reverse engineering of NoSQL property graph databases: The case of Neo4j. *In:* 2017 IEEE International Conference on Big Data (Big Data) (pp. 453–458). IEEE.

[32] Soni, D., T. Ghanem, B. Gomaa and J. Schommer. 2019, June. Leveraging Twitter and Neo4j to study the public use of opioids in the USA. *In:* Proceedings of the 2nd Joint International Workshop on Graph Data Management Experiences & Systems (GRADES) and Network Data Analytics (NDA) (pp. 1–5).

[33] Dharmawan, I.N.P.W. and R. Sarno. 2017, October. Book recommendation using Neo4j graph database in BibTeX book metadata. *In:* 2017 3rd International Conference on Science in Information Technology (ICSITech) (pp. 47–52). IEEE.

[34] Dietze, F., J. Karoff, A.C. Valdez, M. Ziefle, C. Greven and U. Schroeder. 2016, August. An open-source object-graph-mapping framework for Neo4j and Scala: Renesca. *In:* International Conference on Availability, Reliability, and Security (pp. 204–218). Springer, Cham.

[35] Drakopoulos, G. 2016, July. Tensor fusion of social structural and functional analytics over Neo4j. *In:* 2016 7th International Conference on Information, Intelligence, Systems & Applications (IISA) (pp. 1–6). IEEE.

[36] Drakopoulos, G., A. Kanavos, P. Mylonas and S. Sioutas. 2017. Defining and evaluating Twitter influence metrics: A higher-order approach in Neo4j. *Social Network Analysis and Mining*, 7(1): 1–14.

[37] Hölsch, J. and M. Grossniklaus. 2016. An algebra and equivalences to transform graph patterns in Neo4j. *In:* EDBT/ICDT 2016 Workshops: EDBT Workshop on Querying Graph Structured Data (GraphQ).

[38] Hölsch, J., T. Schmidt and M. Grossniklaus (2017). On the performance of analytical and pattern matching graph queries in Neo4j and a relational database. *In:* EDBT/ ICDT 2017 Joint Conference: 6th International Workshop on Querying Graph Structured Data (GraphQ).

[39] Hoksza, D. and J. Jelínek. 2015, September. Using Neo4j for mining protein graphs: A case study. *In:* 2015 26th International Workshop on Database and Expert Systems Applications (DEXA) (pp. 230–234). IEEE.

[40] Johnpaul, C.I. and T. Mathew. 2017, January. A Cypher query based NoSQL data mining on protein datasets using Neo4j graph database. *In:* 2017 4th International Conference on Advanced Computing and Communication Systems (ICACCS) (pp. 1–6). IEEE.

[41] Konno, T., R. Huang, T. Ban and C. Huang. 2017, August. Goods recommendation based on retail knowledge in a Neo4j graph database combined with an inference mechanism implemented in jess. *In:* 2017 IEEE SmartWorld, Ubiquitous Intelligence & Computing, Advanced & Trusted Computed, Scalable Computing & Communications, Cloud & Big Data Computing, Internet of People and Smart City Innovation (SmartWorld/SCALCOM/UIC/ATC/CBDCom/IOP/SCI) (pp. 1–8). IEEE.

[42] Gong, F., Y. Ma, W. Gong, X. Li, C. Li and X. Yuan. 2018. Neo4j graph database realizes efficient storage performance of oilfield ontology. *PloS One*, 13(11): e0207595.

[43] Stark, B., C. Knahl, M. Aydin, M. Samarah and K.O. Elish. 2017, September. Betterchoice: A migraine drug recommendation system based on Neo4j. *In:* 2017 2nd IEEE International Conference on Computational Intelligence and Applications (ICCIA) (pp. 382–386). IEEE.

[44] Wita, R., K. Bubphachuen and J. Chawachat 2017, November. Content-based filtering recommendation in abstract search using Neo4j. *In:* 2017 21st International Computer Science and Engineering Conference (ICSEC) (pp. 1–5). IEEE.

A Comparative Survey of Graph Databases and Software for Social Network Analytics: The Link Prediction Perspective

Nikos Kanakaris[1], Dimitrios Michail[2] and Iraklis Varlamis[2]

[1] University of Patras, Patras, Greece

[2] Harokopio University of Athens, Athens, Greece

In recent years, we have witnessed an excessive increase in the amounts of data available on the Web. These data originate mostly from social media applications or social networks and thus they are highly connected. Graph databases are capable of managing these data successfully since they are particularly designed for storing, retrieving, and searching data that is rich in relationships. This chapter aims to provide a detailed literature review of the existing graph databases and software libraries suitable for performing common social network analytic tasks. In addition, a classification of these graph technologies is also proposed, taking into consideration: (i) the provided means of storing, importing, exporting, and querying data, (ii) the available algorithms, (iii) the ability to deal with big social graphs, and (iv) the CPU and memory usage of each one of the reported technologies. To compare and evaluate the different graph technologies, we conduct experiments related to the link prediction problem on datasets with diverse amounts of data.

2.1 Introduction

Social Network Analysis (SNA) is of great organic value to businesses and society. It encompasses techniques and methods for analyzing the continuous flow of information in offline networks (e.g. networks of employees in labor markets and networks of collaborators in product markets) and online networks (e.g.

Facebook posts, Twitter feeds, and Google maps check-ins) in order to identify patterns of dissemination of information or locate nodes and edges of interest for an analyst.

The importance of graphs for social media analysis was first highlighted by Jacob Moreno. In particular, in his book entitled "Who shall survive?", he attempted to portray the entire population of New York with a network form of relationships [1]. Social networks, like many other real-world phenomena, can be modeled using graphs. A graph is a collection of nodes and edges between them. In the case of social networks, the nodes correspond to people (or groups of people), while the edges represent social or human relationships between the people (or groups).

Consequently, graph processing has become an important part of many social network analytics applications. As social networks evolve and grow in size, the respective social graphs also grow and contain up to trillions of edges. In many cases, these graphs change dynamically over time and are enriched with various metadata. All these increased requirements gave rise to *graph database systems*, which are specialized in storing, processing, and analyzing huge graphs. Graph database systems utilize graph formats to represent data, including nodes, edges, and their properties. The majority of them (e.g. Neo4j, GraphDB, FlockdB, InfiniteGraph) support graph storage and querying. They also support different database models, which either represent data as directional graphs (e.g. as node and edge collections) or as general graph structures that allow data manipulation through a predefined library of graph algorithms.

In this chapter, we perform an overview of the graph-based solutions for social network analytics (SNA). The main contributions of this work are:

- We provide the research community with a comprehensive taxonomy of graph databases and software for graph analysis and mining for social network analytics.
- We perform a comparative evaluation of a selected list of graph mining solutions, aiming to highlight the advantages and disadvantages of each approach.
- We provide the research community with a guideline for choosing the appropriate software solution for each task, taking into account the size and other properties of the dataset under consideration.

The remainder of this chapter is organized as follows. Section 2.2 provides the taxonomy of the graph-based approaches for SNA. Section 2.3 describes a simple SNA task, namely link prediction, and shows how it could be tackled using different popular tools in each category. Section 2.4 performs a comparative evaluation. Using datasets of increasing size, we evaluate the performance and challenge the limits of each solution. Finally, Section 2.5 discusses the advantages and disadvantages of each technology, guides the reader to the selection of the most appropriate tool for each task, and provides directions for future work in the field.

2.2   A taxonomy of graph database approaches for social network analytics

Graph databases organize data either as collections of nodes and relationships between them (edges), where both have a rich set of properties (labeled-property graph model – LPG) or as collections of triples, consisting of a start node, an end node, and an arc for the predicate that describes the relationship between nodes (RDF triple store model). In the latter case, nodes may represent literal values or real-world entities identified from a Uniform Resource Identifier (URI), and the predicates on the arcs are selected from domain-specific ontologies and may also be identified from a URI, whereas key-value pairs are employed in the former case for storing the properties of a node or relationship [2, 3].

The increasing volume of social media data, the dynamics they present, and the often ad-hoc graph processing needs they impose create unique challenges for designing the underlying database systems. To better understand this emerging sector, it is important to have a complete picture of the existing graph database systems. The authors in [4] provide a comprehensive survey of Graph Databases, their models, and their relation with the various back-end technologies they use and adapt in order to store graph data. For this reason, in the following, we examine graph databases under the prism of the analysis of social networks and of the processing needs they impose. So, we proceed with the presentation of the main system categories with emphasis on the storage type, the models, and structures they use to organize graph data in an efficient manner for handling social network analysis tasks.

The tasks of social network analysis comprise, among others, the extraction of useful node-level statistics (e.g. degree, betweenness, and other centrality scores), the computation of network-level statistical measures (e.g. diameter and other distances, average degree, density), computations on the ego network of selected nodes, the analysis of links for the prediction of new links and the detection of communities [5]. In order to support the aforementioned tasks, graph databases provide query capabilities that capitalize on the persistent storage of relationships and avoid the overhead of joins, which is inherent in relational databases [6]. Graph database query languages, such as Neo4j's Cypher [7] provide specialized functions for finding paths, communities, etc, which are fine-tuned to operate on the graph structures.

Graph databases demonstrate many capabilities and advantages in comparison to relational databases, which can be summarized in optimized search capabilities, better performance in data mining tasks, increased storage capacity and better scalability [8]. However, they have limitations related to the maximum number of nodes, relationships, and properties they can handle, the support for sharding into smaller databases, which allows for horizontal scaling to the huge size of social graphs, etc. [9]. To combat these limitations and cover the need for visualizing the graphs and interacting with them, specialized software libraries with more advanced algorithmic capabilities are utilized when the capabilities or performance of a graph database is not adequate.

In order to provide a better understanding of the available solutions in the subsections that follow, we discuss in more depth the various graph database technologies and the software that can be employed for graph-based analysis of social network data. Table 2.1 provides the main software solutions of each type.

2.2.1 Disk-based databases

The need to process big graphs has early raised the interest for distributed data processing frameworks, such as Hadoop MapReduce and Hadoop Distributed File System (HDFS), and specialized extensions or implementations for general-purpose graph processing, such as the PEGASUS library[1] for petascale graph mining operations [10] (PageRank, random walk with restart, graph diameter computation, graph components mining), Apache Giraph [11] and the Apache Hama Bulk Synchronous Parallel (BSP) framework that employs MapReduce for matrix operations [12] (e.g. big matrix multiplications, shortest paths, PageRank) or Pregel [13], which is also based on BSP and Ram-Based System in order to support graph operations (shortest path, PageRank, etc.).

JanusGraph is another graph database that supports several back-end solutions for data storage, such as Apache Cassandra, Apache HBase, Google Cloud Bigtable, Oracle BerkeleyDB, etc. It supports Gremlin query language (and other components of the Apache TinkerPop stack) and can also integrate search of Apache Solr, Apache Lucene, or Elasticsearch applications. IBM, one of the sponsors of the JanusGraph program, is offering a hosted version of JanusGraph on the IBM Cloud called Compose for JanusGraph.

AllegroGraph[2] and OpenLink Virtuoso[3] are two graph databases that have been used for handling Semantic Web data since they support many of the related standards (RDF/S, SPARQL, OWL, etc.). AllegroGraph combines documents and graph databases and provides some features for Social Network Analysis (degree and density computation, degree/closeness and betweenness centrality scores for node and groups, clique detection, etc.). Virtuoso, on the other hand, is a hybrid database that supports SQL and/or SPARQL queries and provides reasoning & inference capabilities.

ArangoDB is an open-source native multi-model database, written in C++, which supports key-value, document and a property graph database model. In ArangoDB the vertices and edges of a graph can be JSON documents. The Arango query language (AQL) is a structured query language that allows to insert and modify data, as well as to query the edge and vertex complex data by filtering, joining or aggregating nodes. The distributed graph processing module, which is on Pregel, supports the execution of popular social network analysis algorithms

---

[1] http://www.cs.cmu.edu/~pegasus/
[2] http://www.franz.com/agraph/allegrograph/
[3] https://virtuoso.openlinksw.com/

Table 2.1: A summary of the main software solutions

Disk-based databases		
Software	**Features**	**Query language**
Janus Graph	Supports several back-end solutions for data storage, such as Apache Cassandra, Apache HBase, Google Cloud Bigtable, Oracle Berkeley DB	Gremlin query language
Allegro Graph	Handling Semantic Web data, since they support many of the related standards (RDF/S, SPARQL, OWL, etc.)	SPARQL
Open Link Virtuoso	A hybrid database that supports SQL and/or SPARQL queries and provides reasoning & inference capabilities	SPARQL
Arango DB	Supports key-value, document and graph database, nodes and edges are JSON documents	AQL
Neo4j	Provides graph traversal operations and full ACID transaction support	Cypher
In-memory databases		
Apache Jena	Offers RDF storage and querying, offers integration with SQL	SPARQL
Sesame	Stores RDF documents	SeRQL
GEMS	HPC solution for big data processing	SPARQL
Memgraph	Supports data replication, ACID transactions and graph querying	Cypher
Redis Graph	Employs the GraphBLAS library for sparse matrix operations on large graphs	Cypher
In-memory libraries		
Software	**Features**	**Language written**
igraph	Offers a sparse graph implementation, and binding for R and Python	C
Boost Graph Libraries	Several graph representations and graph algorithms, supports a distributed-memory model of execution	C++
JUNG	Basic algorithms such as shortest path, and centrality metrics, and a graph drawing component	Java
JGraphT	Supports algorithms for (path) planning, routing, network analysis, combinatorial optimization, applications in computational biology	Java
Google Guava	Provides methods for traversing graphs and getting successors and predecessors of certain nodes	Java
NetworkX	Many algorithms for network analysis	Python
Networkit	Large-scale network analysis using parallelization	C++
SNAP	Single big-memory machine	C++
Webgraph	Combines data compression and algorithms for efficient graph processing	Java

Azure Cosmos DB	Cloud database combines multiple types of databases including graph databases	Gremlin
Orient DB	Optimized for data consistency and low data complexity	SQL

on graphs that are stored in ArangoDB, such as PageRank, Vertex Centrality, Vertex Closeness, Connected Components, Community Detection.

Neo4j[4] is a disk-based transactional graph database implemented in Java, that provides graph traversal operations and full ACID transaction support. It implements the property graph model and stores different graph aspects (nodes, relations, properties) in separate files. It has its own query language (Cypher) that resembles SQL and allows selections, insertions, deletions, and updates to the graph stored in the database. It supports native graph processing through index-free adjacency, i.e. that the nodes keep direct references to their neighboring nodes in physical memory. Thus each node functions as a small pointer to its neighboring nodes, and query times are independent of the overall graph size. Through Cypher queries, Neo4j supports several social network analysis tasks, such as coupling and co-citation counting, retrieving all or selected node properties, etc.

2.2.2   In-memory databases

The need for improving the processing of queries in conjunction with the increased availability of memory storage led to the development of in-memory graph databases.

Apache Jena[5] and Sesame[6] are among the first libraries that offered in-memory RDF storage and querying, while also supporting integration with SQL backends. Graph Database Backend GEMS [14] is a High Performance Computing (HPC) solution for big graph processing. It implements a subset of SPARQL and implements customized extensions that optimize performance in big graphs.

Memgraph[7] is an in-memory graph database that comes as a single node or fully distributed, in order to provide high scalability. It supports data replication, through the RAFT algorithm, Atomicity, Consistency, Isolation, and Durability (ACID) transactions and graph querying using Cypher. In addition, Memgraph implements core graph algorithms, such as Breadth First Search (BFS) and shortest path.

Redis [15] is open-source, in-memory storage that can be used as a database, cache, and/or message broker. It stores data in a key-value format and is a type

---

4   http://neo4j.org/
5   https://jena.apache.org
6   http://rdf4j.org
7   https://memgraph.com/

of NoSQL database. RedisGraph[8] is actually a graph module built into Redis, and since Redis is an in-memory data storage, RedisGraph also stores graphs in memory, thus having high-performance in querying (using Cypher) and indexing. It stores graphs as adjacency matrices using the sparse matrix representation and queries are implemented using linear algebra. The GraphBLAS [16] is a highly optimized library for sparse matrix operations that allows RedisGraph to efficiently perform a wide range of graph analytics. RedisGraph is also based on the Property Graph Model.

2.2.3   In-memory libraries

In-memory graph libraries, with algorithmic support for network analysis, are available in most programming languages.

The igraph[9] [17] library, written in C, contains several optimized algorithms for network analysis combined with an efficient sparse graph implementation. Through bindings, it can also be used in more high-level environments such as R or Python. The Boost Graph Library [18] is an efficient C++ solution that includes several graph representations and graph algorithms. Using templates and generic programming, it allows the user to adjust the underlying graph implementation. Additionally, it includes support for executing several graph algorithms, useful for network analysis, in a distributed-memory model of execution.

The Java Universal Network/Graph Framework (JUNG) library [19] is a free, open-source software library that provides the users with a graph data structure, several basic algorithms such as shortest path, and centrality metrics, and a graph drawing component. JUNG provides a number of graph layout algorithms but also provides algorithms for graph analysis (e.g. clustering and decomposition) and graph metrics (e.g. node centrality, distances, and flows).

JGraphT[10] [20] is a graph processing library, written in Java, that provides generic graph data structures and a sizeable collection of graph analysis algorithms. Although it was primarily intended as a scientific package containing graph-theoretical algorithms, it also supports algorithms for (path) planning, routing, network analysis, combinatorial optimization, applications in computational biology, etc.

Google Guava[11] [21] is another open-source project in Java that contains a list of graph data structures including Graph, ValueGraph, and Network. It also provides methods for traversing graphs and getting successors and predecessors of certain nodes. By special adapter classes, Guava graph implementations can be combined with all the available graph algorithms in JGraphT.

---

[8] https://oss.redislabs.com/redisgraph/
[9] https://igraph.org/
[10] https://jgrapht.org/
[11] https://github.com/google/guava

NetworkX[12] [22] is written in Python and contains graph implementations as well as a large number of algorithms, including many algorithms for network analysis.

More specialized frameworks have also been developed. For example, NetworKit [23] is an open-source package for large-scale network analysis. It is written in C++, employing parallelization when appropriate, and provides Python bindings for ease of use. The Stanford Network Analysis Platform (SNAP) [24] is a high-performance system for the analysis and manipulation of large networks. It focused on the idea of a single big-memory machine and provides a large number of graph algorithms among which some are able to handle dynamic graphs (graphs evolving over time). It is written in C++ with Python bindings.

Webgraph[13] is a framework for the manipulation of large web graphs, which combines data compression and algorithms, such as referentiation and intervalization, for efficient graph processing [25]. It allows users to store a graph in a compact and concise representation either on disk or in-memory and executes algorithms directly using this representation.

2.2.4   Hybrid systems

The Azure Cosmos DB[14] cloud database combines multiple types of databases - conventional tables, document-oriented, column, and graph databases. Its graph database uses the Gremlin query language and the graphical query Application Programming Interface (API). Furthermore, it supports the Gremlin console created for Apache TinkerPop as an alternative interface. Indexing, scaling and geolocation are automatically handled by the Azure cloud, thus offering a useful combination of flexibility and scale [26].

In a similar manner, OrientDB[15] combines different types of data models, such as graph-based, document-based, key-value, and object-based storage. However, all the relationships are stored using the graph model. Similar to Neo4j, OrientDB is written in Java but it does not support Cypher. It is optimized for data consistency and low data complexity.

2.2.5   Graph processing models

In general, graph databases either implement the Vertex-Centric Message Passing logic of the Pregel framework or use the matrix representation (adjacency matrix) and the matrix algebra (e.g. PEGASUS, GBASE, etc). Thus, they rely on MapReduce or Spark for distributed execution [27]. There is also an array of approaches that employs a subgraph-centric programming model, which operates

---

[12] https://networkx.org/
[13] https://webgraph.di.unimi.it/
[14] https://azure.microsoft.com/en-us/services/cosmos-db/
[15] https://www.orientdb.org/

on small subsets of the input graph instead of individual edges or vertices. As a result, they can solve local neighbor tasks (e.g. local clustering coefficient computation) in a more efficient manner [28].

2.3   A social network analysis case study and the role of graph databases

When it comes to social network analysis (SNA), several graph database software that have been primarily designed for graph querying show limited capabilities and have restrictions that do not allow them for properly performing on large social graphs [29]. The so-called big graphs, from social networks such as Facebook, Twitter, etc. have billions of nodes and hundreds of billions of edges, and because of this size, they need special algorithms for data compression, sharding, and processing in order to efficiently be stored and managed. In contrast to Resource Description Framework (RDF) processing, SNA depends more on high read performance and batch mode analytics than on write performance and CRUD operations (Create, Read, Update, Delete) or querying.

In this section, we describe a basic link prediction approach and select three different technologies in order to implement it, representing different approaches that a user might take when faced with SNA on big graphs.

2.3.1   The link prediction problem

Link prediction in a graph is a fundamental problem in the wider graph study area and aims to assess the likelihood of an edge (link) based on some form of graph observed during its "life" (snapshot). In the case of social network analysis, edges/ links correspond to the interactions between social users, and as social networks become more and more popular, several techniques have been developed that aim to predict short-term user interactions. Jaccard coefficient, common neighbor, and the algorithm of Adamic and Adar [30, 31, 32] are some of the alternative algorithms that are available in the SNA literature and have been implemented in most graph databases and related libraries.

Since the prediction of an unknown edge is not easy to be validated, what SNA algorithms usually do is to assume the dynamic nature of social graphs, in which new edges are continuously added or removed, and then take a snapshot of the graph at point t_1 in time and use it for training or analysis in order to predict edges that will probably appear in a second snapshot of the graph at point t_2 in time. This allows to measure the precision and recall of any algorithm and thus be able to compare algorithms for their performance.

There are many algorithms for link prediction [33], such as those that are based on the graph distance metric, which calculate the shortest path between two vertices and associate it with the existence of a direct edge between the two, at a later point in time. Next, is the common neighbor metric, in which the common neighbors between two vertices are calculated and affect the probability of an edge to appear between the two vertices in the future. Essentially, what this metric does is to "close triangles" between not directly connected vertices, which in the social network context assume that the friend of a mutual friend is likely to be our friend in the future.

Another algorithm is that of Adamic & Adar [31] which was introduced for predicting links on social networks (Facebook, Twitter, etc). The algorithm also calculates the common neighbors of two vertices, but in addition to their simple count, it calculates the sum of the inverse logarithm of the degree of each neighbor. Thus, it manages not to reject the scenarios in which low degree nodes play an important role in predicting links. The formula for measuring the Adamic-Adar index for a potential edge (x, y) is given in Eq. 2.1, where $N(u)$ denotes the set of neighbors of node u.

$$A(x, y) = \sum_{\mu \in N(x) \cap N(y)} \frac{1}{\log |N(u)|} \qquad (2.1)$$

2.3.2 A simple link-prediction workload

Given a graph G, the goal is to compute the top-k pair of vertices, which are not already connected by an edge, with the higher link prediction score. For simplicity, we focus only on the Adamic-Adar score and we restrict ourselves to a single multi-core machine setup. As our graphs are social network graphs which are usually sparse and vertex degrees exhibit a power-law distribution, we restrict computation of link prediction scores only between vertices that are highly connected. Thus, we assume as input from the user a threshold d_{min} which denotes the minimum degree that a vertex should have in order to be included in a candidate pair. The experiment can be described using the following steps:

1. Load the graph
2. Find vertices which have degree $\geq d_{min}$ and compute pair of vertices with degree $\geq d_{min}$ which are not already connected
3. Split candidate pairs into separate lists based on the number of cores of the machine
4. Compute Adamic-Adar for each list of candidate pairs in parallel and keep only the top-k
5. Reduce separate top-k results into global top-k results

This simple experimental procedure can be implemented using a variety of different technologies in combination with graph databases.

2.3.3 Approach 1: Native graph database with algorithmic support (Neo4j)

A popular approach for making link prediction is to execute the whole workload inside a graph database. Neo4j provides a large variety of graph algorithms, including community detection, centrality and similarity algorithms, node embeddings, and algorithms for link prediction and pathfinding. Support for all these algorithms is included as a separate library, which is highly coupled with the core system and is aware of the underlying storage implementation. In order to process big graphs that do not fit in the main memory, Neo4j supports partitioning them over multiple servers and aggregating the results of each node. For such purposes, Neo4j provides an integration with the Pregel API for custom Pregel computations.

Since Neo4j contains native support for the Adamic-Adar score, we keep almost all computation inside the database. After loading the graph in the DB, we use a Python-based driver program that executes the necessary steps. Finding the vertices which high degree is performed using the following query:

```
MATCH(m1:Member)-[:Friend]->(m2:Member)
WITH m1, count(m2) as degree, collect(m2)as friends
WHERE degree >={min_degree}RETURN m1, friends
```

Then, after building the candidate pairs in the Python driver and splitting them into independent lists depending on the number of cores of the machine, we execute the following query for each candidate pair:

```
MATCH (m1:Member {name: '%s'})
MATCH (m2:Member {name: '%s'})
RETURN gds.alpha.linkprediction.adamicAdar(m1, m2,
        {relationshipQuery:'Friend', orientation:'NATURAL'})
```

The final top-k computation is performed in the Python driver.

2.3.4   Approach 2: Sparse matrices in-memory graph database (RedisGraph)

RedisGraph uses sparse matrices in order to store graph structure in memory. The use of sparse representation allows RedisGraph to handle large graphs and perfectly fits the sparsity of social networks. Using matrices as its data model and linear algebra to implement the various operations, RedisGraph achieves high performance on sparse graphs but also changes the way popular graph mining algorithms are implemented [16]. RedisGraph incorporates the SuiteSparse implementation of the GraphBLAS specification, which allows high shared-memory parallelism, by recurring to OpenMP. GraphBLAS defines new data types, such as monoids, semirings, masks, and descriptors, which are combined with vectors and matrices in order to implement the various graph algorithms [34]. Examples of available algorithms are Breadth-First Search for graph traversal [16], betweenness centrality, PageRank, Single-Source shortest paths, triangle counting, and connected components [35].

RedisGraph does not contain native support for the Adamic-Adar score. Thus, in this case, the score is calculated by our Python driver program. After loading the graph, the following query is executed which returns the relevant vertices along with their neighbors.

```
MATCH (m1:Member)-[:Friend]->(m2:Member)
WITH m1, count(m2) as degree, collect(m2) as friends
WHERE degree >= {min_degree} RETURN m1, degree, friends
```

The computation of all candidate pairs and their split into multiple separate lists for parallel processing is performed, as previously, in the Python driver program. The final computation of Adamic-Adar is performed using Python. One complication which arises here, due to the nature of the Adamic-Adar score, is the need for degrees of vertices that are not in candidate pairs. When such vertices are encountered, an additional query for vertex v is performed and its result is cached in case it is needed a second time.

```
MATCH (m1:Member {{name:{v}}})-[:Friend]->(m2:Member)
WITH m1, count(m2) as degree RETURN m1, degree
```

The final aggregation of top-k results is performed in Python just like in the Neo4j case.

2.3.5   Approach 3: Sparse matrices graph library (JGraphT)

JGraphT [20] is a in-memory graph library. It is built around a generic Graph<V, E> interface and supports the use of several alternative data representation models. In this case, it is used with a sparse matrix representation which fits nicely to the static nature of the graph, as well as its social network characteristics. The library supports many algorithms such as graph traversals, shortest paths, network flows, and others. In terms of social network analysis, it offers algorithms for clique detection, graph clustering, and link prediction.

Since JGraphT supports the Adamic-Adar score computation, the whole procedure is executed in our driver program. Instead of Python in this case we utilize a Java driver program. The Java driver performs the same steps as in the case of Neo4j or RedisGraph. It first builds the candidate pairs, then splits the lists into multiple separate candidate pair lists and executes the Adamic-Adar computation in parallel. Finally, it aggregates top-k and returns the result. Note that loading the graph is performed in the Java driver, but we do not include the loading time in the results, for a fair comparison with the other approaches.

2.4   Experimental comparison

We ran our experiments on a Linux server (Ubuntu 16.04.7) with an Intel Xeon CPU e5-2680, 16 cores at 2.5GHz, and 196GB of RAM. For the implementation of our experiments, we used the Python and Java programming languages, the JGraphT library [20], the Neo4j and RedisGraph graph databases. The full code of our experiments is available for download at https://github.com/ d-michail/ graphs-social-link-prediction.

2.4.1　Datasets

In order to compare the various approaches, we used four datasets, which are available from the Stanford Network Analysis Project (SNAP)[16]. Each dataset has a different number of nodes and edges. In particular:

- **wiki-Vote**[17]: Wikipedia is a free encyclopedia written collaboratively by volunteers around the world. A small part of Wikipedia contributors are administrators, who are users with access to additional technical features that aid in maintenance. In order for a user to become an administrator, a Request for adminship (RfA) is issued and the Wikipedia community via a public discussion or a vote decides who to promote to adminship. Using the latest complete dump of Wikipedia page edit history (from January 3, 2008) we extracted all administrator elections and voting history data. This gave us 2,794 elections with 103,663 total votes and 7,066 users participating in the elections (either casting a vote or being voted on). Out of these 1,235 elections resulted in a successful promotion, while 1,559 elections did not result in the promotion. About half of the votes in the dataset are from existing admins, while the other half comes from ordinary Wikipedia users. The network contains all the Wikipedia voting data from the inception of Wikipedia till January 2008. Nodes in the network represent Wikipedia users and a directed edge from node i to node j represents that user i voted on user j.
- **soc-Epinions1**[18]: This is a who-trust-whom online social network of a general consumer review site Epinions.com. Members of the site can decide whether to "trust" each other. All the trust relationships interact and form the Web of Trust which is then combined with review ratings to determine which reviews are shown to the user.
- **ego-Twitter**[19]: This dataset consists of 'circles' (or 'lists') from Twitter. Twitter data was crawled from public sources. The dataset includes node features (profiles), circles, and ego networks. Data is also available from Facebook and Google+.
- **soc-LiveJournal1**[20]: LiveJournal is a free online community with almost 10 million members; a significant fraction of these members are highly active. For example, roughly 300,000 update their content in any given 24-hour period. LiveJournal allows members to maintain journals, individual and group blogs, and it allows people to declare which other members can have access to them.

---

[16] http://snap.stanford.edu/index.html

[17] http://snap.stanford.edu/data/wiki-Vote.html

[18] http://snap.stanford.edu/data/soc-Epinions1.html

[19] http://snap.stanford.edu/data/ego-Twitter.html

[20] http://snap.stanford.edu/data/soc-LiveJournal1.html

2.4.2   Evaluation results

For the experimental evaluation, we used the datasets described in Section 2.4.1, the subsets of nodes in each that are depicted in Table 2.2, and an implementation of the Adamic-Adar algorithm in Neo4j, RedisGraph, and JGraphT. This allowed us to compare the three solutions, which span the broad range of graph databases (in terms of disk and memory) and software that can be used for social network analytics, with an increasing data load and test their limits and capabilities.

Table 2.2. Statistics of the datasets

	# nodes	# edges	# nodes ≥ 100	# nodes ≥ 250	# nodes ≥ 500
wiki-Vote	7,115	103,689	231	41	7
soc-Epinions1	75,879	508,837	836	114	21
ego-Twitter	81,306	1,768,149	2992	197	26
soc-LiveJournal1	4,847,571	68,993,773	94,657	15,107	2,691

'# nodes' and '# edges' correspond to the number of nodes and edges of the original dataset. '# nodes 100', '# nodes 250' and '# nodes 500' correspond to the number of nodes that have a degree greater or equal than 100, 250 and 500 respectively

As shown in Tables 2.3 to 2.5 in-memory software solutions such as JGraphT and RedisGraph did a better utilization of the memory than their disk-based graph database alternatives. They used compression in order to improve their performance and thus, allowed to scale up to larger datasets, when the graph fits into the available memory. On the other hand, disk-based graph databases

Table 2.3. Performance results, on all datasets using a minimum degree of 100

	wiki-Vote		soc-Epinions1		ego-Twitter		soc-LiveJournal1	
	Time	Memory	Time	Memory	Time	Memory	Time	Memory
Neo4j	1m 22s	103 GB	58m 39.6s	106 GB	928m 6.6s	112 GB	Out of memory	
RedisGraph	0m 3.2s	890 MB	0m 29s	1.04 GB	4m 12.14s	2.72 GB	Out of memory	
JGraphT	0m 1s	1.3 GB	0m 9s	6.8 GB	1m 1s	40.3 GB	Out of memory	

Table 2.4. Performance results, on all datasets using a minimum degree of 250

	wiki-Vote		soc-Epinions1		ego-Twitter		soc-LiveJournal1	
	Time	Memory	Time	Memory	Time	Memory	Time	Memory
Neo4j	0m 14s	48.6 GB	1m 23.8s	80.1 GB	4m 19.8s	61.1 GB	Out of memory	
RedisGraph	0m 0.8s	860 MB	0m 6.7s	1004 MB	0m 8.9s	1.16 GB	438m 36.5s	40.5 GB
JGraphT	0m 0.9s	1.1 GB	0m 1.9s	2.4 GB	0m 5.4s	4.5 GB	68m 14s	130 GB

Table 2.5. Performance results, on all datasets using a minimum degree of 500

	wiki-Vote		soc-Epinions1		ego-Twitter		soc-LiveJournal1	
	Time	Memory	Time	Memory	Time	Memory	Time	Memory
Neo4j	0m 5s	13.3 GB	0m 13.6s	48.2 GB	0m 5.5s	40 GB	363m 25s	149 GB
RedisGraph	0m 0.4s	860 MB	0m 3s	987 MB	0m 3.5s	1.14 GB	152m 34s	8.66 GB
JGraphT	0m 0.7s	900 MB	0m 1.7s	1.3 GB	0m 4.2s	2.9 GB	9m 38s	97 GB

do not compress data, and thus need to load a large amount of data from disk to memory during the query execution, which introduces significant overhead in query execution. This is the reason that Neo4j ran out of memory for both LiveJournal subsets that used a minimum degree of 100 or 250. JGraphT had similar memory issues with the same dataset and the nodes with a minimum degree of 100, but ran successfully on the fewer nodes that had a minimum degree of 250. RedisGraph outperformed both solutions in terms of memory usage in all cases.

An increasing minimum degree threshold selected fewer and fewer nodes to apply the algorithm and thus, speeded up the execution in all cases. The effect is insignificant for small datasets (e.g. the wiki-Vote), but brought an important decrease in memory usage and execution time for larger datasets. For example, JGraphT in the case of the LiveJournal dataset employed 75% of the memory needed for examining nodes with degrees higher than 250 (i.e. 15107 nodes), when it examined nodes with degrees higher than 500 (i.e., 2,691 nodes).

A comparison of the solutions in terms of the execution time shows that the tasks were executed quite fast for small datasets in all systems. However, even in the smallest dataset, Neo4j was 10 times slower than RedisGraph and JGraphT, whereas on larger graphs the execution time was even worse. In general, JGraphT was faster than RedisGraph in the same graphs and this was most probably due to a more efficient implementation of the algorithm in the former case. When the threshold on minimum degree was high and the resulting nodes are few (e.g., in Table 2.5 RedisGraph was faster than JGraphT), mainly because it handled more efficiently the storage of the compressed graph in memory.

In order to provide a better visual comparison of how the algorithmic implementations performed in each system, in terms of execution time for an increasing number of nodes (depending on the dataset and the minimum degree threshold), we draw the respective plots for each degree threshold. The results are shown in Figures 2.1 to 2.3.

2.5   Discussion and conclusion

In this chapter, we provided a literature review of the existing graph databases and software libraries, which are suitable for performing common social network analytic tasks. Furthermore, a taxonomy of these graph technologies was also

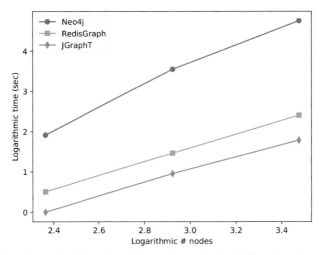

Figure 2.1. Execution times for an increasing graph size (using min. degree = 100).

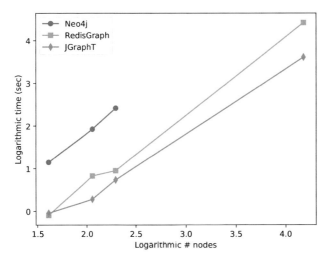

Figure 2.2. Execution times for an increasing graph size (using min. degree = 250).

proposed, taking into consideration the provided means of storing, importing, exporting, and querying data.

Emphasis was given on the memory and disk usage requirements of each software, that affects the scalability of each solution in large social graphs. Also, on the scope of each group of software, which varies between graph databases that focus on query answering and software libraries that emphasize on network analysis algorithms.

The survey of existing graph-based solutions for social network analysis showed that disk-based and memory-based are the two main alternatives from

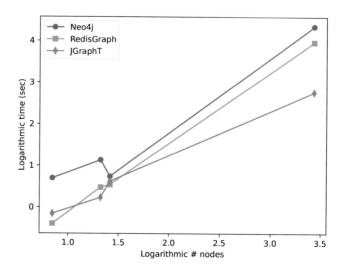

Figure 2.3. Execution times for an increasing graph size (using min. degree = 500).

the graph database side, whereas many specialized graph processing and analysis software exists that implement algorithms for supporting such tasks in memory.

The experimental evaluation results demonstrated the superiority of specialized software for SNA, against graph databases in terms of applying specific algorithms on networks that fit (in compressed format or not) in main memory. Disk-based graph databases are still the preferable solution for custom queries, that target knowledge extraction from the networks. Nice compromises already exist [36] where Cypher queries directly from Neo4j could be converted on the fly in JGraphT graphs in order to execute more sophisticated algorithms.

Graph database software was more consistent in the implementation of database operations and features, such as transaction management and querying, but provided fewer graph analysis functionalities and algorithms. On the other side, the specialized software for social network analysis are mostly oriented in the implementation of a wide range of algorithms and optimizing their performance in the execution of these algorithms. Despite their improved performance in algorithmic tasks, their scalability was limited by the memory available in the system, which in turn restricted the graph size that they could handle.

The social network analysis task examined in this chapter (i.e. link prediction) is a typical task that can be efficiently handled by an in-memory algorithm, but can also be implemented in an offline (i.e. disk-based) setup using a graph database and its query engine. This allowed us to re-write the task for the different types of software solutions and perform an end-to-end comparison, using various graph sizes and complexities and the respective query or algorithmic workloads. With this task at hand it was able to directly compare the solutions in terms of memory and time performance and test their limits. The comparison showed the superiority of in-memory solutions in terms of execution speed. It also showed a more concise memory usage from such systems, which in terms of our experiments allowed them to scale up to larger graphs, that their graph database counterpart.

Future work directions include the evaluation of more system and software alternatives, on a wider range of tasks that comprise the execution of social network mining algorithms of varying complexity and their translation to graph queries that can be directly applied to graph database software.

Appendix

Acronyms used in the chapter is shown in Table 2.6.

Table 2.6. Acronyms used in the chapter

Acronym	Full name
SNA	Social Network Analysis
LPG	Labeled-property graph model
URI	Uniform Resource Identifier
HDFS	Hadoop Distributed File System
HPC	High Performance Computing
ACID	Atomicity, Consistency, Isolation and Durability
BFS	Breadth First Search
JUNG	Java Universal Network/Graph Framework
SNAP	Stanford Network Analysis Platform
API	Application Programming Interface
RDF	Resource Description Framework
CRUD	Create, Read, Update, Delete

References

[1] Moreno, J.L. 1934. Who shall survive? A new approach to the problem of human interrelations. Nervous and Mental Disease Publishing Co.

[2] Angles, R. and C. Gutierrez. 2008. Survey of graph database models. *ACM Computing Surveys (CSUR)*, 40(1): 1–39.

[3] Kaliyar, R.K. 2015. Graph databases: A survey. *In:* International Conference on Computing, Communication & Automation (pp. 785–790). IEEE.

[4] Besta, M., E. Peter, R. Gerstenberger, M. Fischer, M. Podstawski, C. Barthels, G. Alonso and T. Hoefler. 2019. Demystifying graph databases: Analysis and taxonomy of data organization, system designs, and graph queries. arXiv preprint arXiv:1910.09017.

[5] Tabassum, S., F.S.F. Pereira, S. Fernandes and J. Gama. 2018. Social network analysis: An overview. *Wiley Interdisciplinary Reviews: Data Mining and Knowledge Discovery*, 8(5): e1256.

[6] Wood, P.T. 2012. Query languages for graph databases. *ACM Sigmod Record*, 41(1): 50–60.

[7] Francis, N., A. Green, P. Guagliardo, L. Libkin, T. Lindaaker, V. Marsault, S. Plantikow, M. Rydberg, P. Selmer and A. Taylor. 2018. Cypher: An evolving query language for property graphs. *In:* Proceedings of the 2018 International Conference on Management of Data (pp. 1433–1445).

[8] Guia, J., V. Goncalves Soares and J. Bernardino. 2017. Graph databases: Neo4j analysis. *In:* ICEIS (1) (pp. 351–356).

[9] Fernandes, D. and J. Bernardino. 2018. Graph databases comparison: Allegrograph, Arangodb, Infinitegraph, Neo4j, and Orientdb. *In:* DATA (pp. 373–380).

[10] Kang, U., C.E. Tsourakakis and C. Faloutsos. 2011. Pegasus: Mining peta-scale graphs. *Knowledge and Information Systems*, 27(2): 303–325.

[11] Martella, C., R. Shaposhnik, D. Logothetis and S. Harenberg. 2015. Practical graph analytics with apache giraph, volume 1. Springer.

[12] Siddique, K., Z. Akhtar, E.J. Yoon, Y.-S. Jeong, D. Dasgupta and Y. Kim. 2016. Apache hama: An emerging bulk synchronous parallel computing framework for big data applications. *IEEE Access*, 4: 8879–8887.

[13] Malewicz, G., M.H. Austern, A.J.C. Bik, J.C. Dehnert, I. Horn, N. Leiser and G. Czajkowski. 2010. Pregel: A system for large-scale graph processing. *In:* Proceedings of the 2010 ACM SIGMOD International Conference on Management of Data (pp. 135–146).

[14] Castellana, V.G., A. Morari, J. Weaver, A. Tumeo, D. Haglin, O. Villa and J. Feo. 2015. In-memory graph databases for web-scale data. *Computer*, 48(3): 24–35.

[15] Da Silva, M.D. and H.L. Tavares. 2015. Redis Essentials. Packt Publishing Ltd.

[16] Davis, T.A. 2019. Algorithm 1000: Suitesparse: Graphblas: Graph algorithms in the language of sparse linear algebra. *ACM Transactions on Mathematical Software (TOMS)*, 45(4): 1–25.

[17] Csardi, G. and T. Nepusz. 2006. The igraph software package for complex network research. *InterJournal, Complex Systems*, 1695(5): 1–9.

[18] Siek, J.G., L-Q. Lee and A. Lumsdaine. 2001. Boost Graph Library: User Guide and Reference Manual, The Pearson Education.

[19] O'Madadhain, J., D. Fisher, S. White and Y. Boey. 2003. The JUNG (java universal network/graph) framework. University of California, Irvine, California.

[20] Michail, D., J. Kinable, B. Naveh and J.V. Sichi. 2020. Jgrapht—A java library for graph data structures and algorithms. *ACM Transactions on Mathematical Software (TOMS)*, 46(2): 1–29.

[21] Bejeck, B. 2013. Getting Started with Google Guava. Packt Publishing Ltd.

[22] Hagberg, A., P. Swart and D.S. Chult. 2008. Exploring network structure, dynamics, and function using networkx. Technical Report, Los Alamos National Lab. (LANL), Los Alamos, NM (United States).

[23] Staudt, C.L., A. Sazonovs and H. Meyerhenke. 2016. Networkit: A tool suite for large-scale complex network analysis. *Network Science*, 4(4): 508–530.

[24] Leskovec, J. and R. Sosič. 2016. Snap: A general-purpose network analysis and graph-mining library. *ACM Transactions on Intelligent Systems and Technology (TIST)*, 8(1): 1.

[25] Boldi, P. and S. Vigna. 2004. The webgraph framework I: Compression techniques. *In:* Proceedings of the 13th International Conference on World Wide Web (pp. 595–602).

[26] Paz, J.R.G. 2018. Introduction to Azure Cosmos DB. *In:* Microsoft Azure Cosmos DB Revealed (pp. 1–23). Springer.

[27] Yan, D., Y. Tian and J. Cheng. 2017. Systems for Big Graph Analytics. Springer.

[28] Yan, D., Y. Bu, Y. Tian and A. Deshpande. 2017. Big graph analytics platforms. *Foundations and Trends in Databases*, 7(1-2): 1–195.

[29] Pokorný, J. 2015. Graph databases: Their power and limitations. *In:* IFIP International Conference on Computer Information Systems and Industrial Management (pp. 58–69). Springer.

[30] Zhou, T., L. Lü and Y.-C. Zhang. 2009. Predicting missing links via local information. *The European Physical Journal B*, 71(4): 623–630.

[31] Adamic, L.A. and E. Adar. 2003. Friends and neighbors on the web. *Social Networks*, 25(3): 211–230.

[32] Liben-Nowell, D. and J. Kleinberg. 2007. The link-prediction problem for social networks. *Journal of the American Society for Information Science and Technology*, 58(7): 1019–1031.

[33] Yuliansyah, H., Z.A. Othman and A.A. Bakar. 2020. Taxonomy of link prediction for social network analysis: A review. IEEE Access.

[34] Buluc, A., T. Mattson, S. McMillan, J. Moreira and C. Yang. 2017. The Graphblas C API specification. *Tech. Rep*, 88. GraphBLAS.org

[35] Szárnyas, G., D.A. Bader, T.A. Davis, J. Kitchen, T.G. Mattson, S. McMillan and E. Welch. 2021. Lagraph: Linear algebra, network analysis libraries, and the study of graph algorithms. arXiv preprint arXiv:2104.01661.

[36] JGraphT Neo4j client. 2021. https://github.com/murygin/jgrapht-neo4j-client. Accessed: 2021-06-01.

A Survey on Neo4j Use Cases in Social Media: Exposing New Capabilities for Knowledge Extraction

Paraskevas Koukaras [0000-0002-1183-9878]

The Data Mining and Analytics Research Group, School of Science and Technology, International Hellenic University, GR-570 01 Thermi, Greece
e-mail: p.koukaras@ihu.edu.gr

Neo4j is the leading NoSQL graph database. In the past few years, it has gained popularity with ever more businesses and enterprises using it. In addition, many universities have incorporated teaching Neo4j within their syllabi. At the same time, many organizations were forced to change and trust NoSQL databases due to the characteristics of relational databases, such as cumbersomeness and the inability to adapt to the needs of data volume eruption and the incorporation of new and diverse data types. This chapter provides a literature survey and analyzes major use cases and applications of Neo4j, focusing on the domain of social media. At the same time, it categorizes them according to identified context. The categorization includes topic detection/extraction, recommendation systems, branding/marketing, learning environments, healthcare analytics, influence detection, fake/controversial information, information modeling, environment/disaster management, profiling/criminality, attractions/tourism, and metrics/altmetrics. For each category, representative examples are showcased elevating the necessity and benefits of using a NoSQL database such as Neo4j for various applications.

3.1 Introduction

Neo4j[1] is a graph database designed from the bottom up to use both data and data connections. Neo4j links data as it is stored, allowing queries never seen before, at

---

1 https://neo4j.com/

speeds envisaged. Neo4j has a flexible structure defined by recorded connections between data entries, unlike conventional databases, which arrange data in rows, columns, and tables. Each data record, or node, in Neo4j, holds direct links to all the nodes to which it is linked. Neo4j runs queries in highly connected data quicker and with more depth than other databases since it is built around this basic yet effective optimization.

This chapter performs a literature review and gathers the most important use cases of Neo4j in the domain of Social Media (SM). It also offers a categorization of these use cases intending to highlight the benefits of NoSQL databases. The literature survey reviews publications in the field of SM, where Neo4j was employed. More than 100 research articles were found which were shortlisted having evaluated their content, publication year and relevance with the SM domain. Having performed this pre-analysis this chapter reports on 40 articles and their context categorization.

Telecommunications, financial services, logistics, hotels, and healthcare are among the areas that benefit from graph databases. Graph databases can be employed efficiently in several fields, including Social Network (SN) analysis, data management and more. Graph databases are the fastest-growing category in Database Management Systems (DBMS), with more than 25% of businesses using them since 2017. The primary features and benefits of graph databases are the following [1]:

- They support new data types and data storage in the magnitude of petabytes.
- Graphs are a natural type of information representation, thus they are quite intuitive.
- They offer high-performance deep querying when compared with relational databases.
- The information search is considerably more improved than in relational databases because it makes use of proximity data from one or more graph database roots (primary nodes).
- They are optimized for all data mining tasks expanding the capabilities of SQL databases.
- They are ideal for data interconnection that represents real-world semantics.
- They are particularly flexible in development, since they may be readily modified over time, either by adding or removing information.

Graph databases, which incorporate these features, are used by organizations such as eBay to improve the output of their recommendation systems, which provide real-time user-centric product suggestions [2].

Neo4j is the world's most popular graph database, having the largest reference area and broad usage[2]. It is widely considered the industry standard in this sector and is widely utilized in health care, government, automobile production, military, and other industries. Neo4j is a free and open-source project written in the Java programming language. According to its designers, it is a completely transactional

---

2 https://db-engines.com/en/ranking_trend

database with a persistent Java engine that allows structures to be saved as graphs rather than tuples.

Neo4j was first released in 2007 and is divided into three product releases: Community, Government, and Enterprise. At the time of writing, the Community Edition is the trial edition, which includes all fundamental capabilities. For a month, the Enterprise Edition grants access to the full edition of Neo4j. The Government Edition is an update to the Enterprise edition that focuses on government services. The primary distinctions between Community and Enterprise/Government Editions include a sophisticated monitoring system, increased database scalability, robust management of database locks, the presence of online backup, a high-performance level of memory cache, and more [2].

The remainder of this chapter is structured as follows. Section 2 reports on the state-of-the-art data analytics in the domain of SM that utilize Neo4j. Section 3 analyzes and outlines the resulting categorization of the state-of-the-art, which is the main goal of this work. Finally, section 4 discusses the context of this chapter, the implications of such a taxonomy, as well as future directions resulting from the conducted survey on the state-of-the-art Neo4j use cases in SM.

3.2   Literature review

This section surveys various case studies on Neo4j utilization in SM. These are categorized into respective sections based on the context of each investigated study. The resulting categories are topic detection and extraction, recommendation systems, branding and marketing, learning environments, healthcare analytics, influence detection, fake and controversial information, information modeling, environmental and disaster management, profiling and criminality, attractions and tourism, and finally metrics and altmetrics. The goal is to support the conception of a taxonomy of Neo4j recent use cases that result from the exploitation of data from SM.

3.2.1   Topic detection and extraction

Real-time short messages with great data scale and a variety of topics are of great scientific interest. These data characteristics in SNs have led to the conception of novel algorithms and techniques for SM analytics. Hot topic detection and event tracking mostly rely on Natural Language Processing (NLP). Although unsupervised data mining methods may be considered language-independent, there are issues such as noisy sentences, grammar mistakes, and user-invented words. Also, the semantic relationships of words and especially synonyms are in most cases ignored.

Therefore, a novel approach utilizes transformers along with an incremental community detection algorithm. Transformers calculate the semantic relation of words, while a proposed graph mining technique defines topics integrating simple structural rules. Multimodal data, image and text, label entities, and new enhanced topics are extracted. For validating this approach NoSQL technologies are utilized. More specifically, MongoDB and Neo4j are combined. Such a system is validated by three different datasets reporting on higher precision and recall when compared with other state-of-the-art systems [3].

3.2.2 Recommendation systems

Recommendation systems expose multiple business applications for enterprise decisions. Historical data, such as sales or customer feedback is very important since it reflects the image of the organization to its customers. A recommendation model attempts to measure the product-to-product influence and in case a user purchases an influential product, the recommender system can adjust and increase its recommendation weight and promote it to other customers. To that end, association rule mining is incorporated by conceiving and proposing a new technique that yields a better performance from the Apriori algorithm. The developed framework utilizes Neo4j for data modeling and is being validated in a real-world dataset from Amazon customer feedback [4].

Graph databases are ideal for storing information that matches the human-like perception in terms of data modeling (objects and relations), making it easier to extract knowledge. A two-layer knowledge graph database is proposed in [5]. In this approach, there is a concept layer and an instance layer. The first is the graph representation of an ontology representation. The second is the data associated with the concept nodes. This model is validated in a retail business transaction dataset through extensive querying and evaluation of results. It is implemented in Neo4j and Jess's reasoning engine. Knowledge graph structures are being used for presenting the recommendation results while the performance is evaluated in time efficiency and the novelty of recommendations.

Graph structures are self-explanatory and flexible in data storage and presentation. A novel approach suggests to people the shortest path for going shopping based on their historic shopping behavior. The novelty resides in the fact that it considers the shortest path to the nearest mall utilizing QGIS[3]. Neo4j stores and reviews various categories of products. A variety of recommendation systems is also presented attempting to offer guidelines for utilizing a graph database for developing a recommendation system while reporting some benefits and drawbacks of common recommendation algorithms [6].

3.2.3 Branding and marketing

Online marketing is utilized to establish a brand and to increase its popularity and online presence. Advertisements present the services and products of a company and are necessary for an effective online marketing strategy. A novel brand analysis framework consists of three components. (i) A website that fosters specific queries that return webpage results from search engines such as Google, Yahoo and Bing. (ii) A content scraping framework that crawls webpage form results. (iii) Ranking of the retrieved webpages with text edge processing as a

---

3 https://qgis.org/en/site/

metric of interest and occurrence of results related to the initial search query. Sentiment analysis (SA) is also utilized to report on the positive/negative impact of text and its respective terms on the webpage. A final rank is applied to the retrieved web pages and SN pages. The goal is to identify the appropriate online location for advertising the brand yielding a positive impact and high affinity with the ideal content. This approach constitutes a targeted brand analysis that enables benefits for both customers and advertisement agencies [7].

Online market research strategies involve the dissemination of IT product information to retrieve customer input. Yet, SM customer responses tend to be unstructured and very large in volume. An automated approach for customer opinion extraction would wield great benefits for the effectiveness of business ads. An intelligent method measures the inter-profile causality, value structures, and user attitudes considering replies in SM, such as YouTube. By utilizing the media/information richness theory to report on the agility and information richness features. Consequently, a deep SA approach is proposed that seems to outperform other legacy approaches [8].

Customer review websites and SM are great sources for mining product feature data. Most design methodologies for product feature extraction assume that the product features expressed by customers are clearly stated and comprehended and can be mined straight ahead. Although, that is very rarely accurate. A novel inference model assigns the most probable explicit product feature that the customer desires based on an implicit preferences expression in an automated way. The algorithm adjusts its inference functionality by hypothesizing and utilizing ground truth. This approach is validated through its statistical evaluation in a case study of smartphone product features utilizing a Twitter dataset [9].

3.2.4 Learning environments

Social learning applications aim to boost the chances of grouping students, teachers, courses, and study groups all in one place within the context of a university. Yet, its effectiveness is highly dependent on the level of successful interactivity and relationships between all the involved parties. The necessary information modeling and retrieval for such a scheme builds upon the social learning graphs' formalism. To that end, Neo4j is utilized as a graph database and a web application is implemented with Java and Spring.

The use case for validating the proposed testing scheme is the University of Craiova which utilizes the Tesys and MedLearn e-learning platforms. Combining the experience of SM with an e-learning environment should yield benefits (responsiveness, familiarity, attraction) since SNs have higher user penetration among young people. That way, education can be a process that is secure, and transparent while integrating the fun of SM for enhancing an academic experience [10].

3.2.5   Healthcare analytics

Identifying and evaluating patient insights from SM allows healthcare providers to have a better understanding of what patients desire and identify as their trouble areas. Healthcare organizations cannot ignore the importance of monitoring and analyzing popular SM platforms such as Twitter and Facebook. A healthcare provider must be able to successfully connect with their patients and respond to their desires.

To begin, it is necessary to define the connections between the contents of the healthcare pain points as reflected by the SM discussion in terms of a sociographic network, in which the elements forming these dialogues are characterized as nodes and their interactions as links. Conversation communities will be represented by discussion groups of nodes that are well-linked.

Several consumer-concealed insights can be inferred by identifying these conversation communities. For example, by using techniques such as visualizing conversation graphs relevant to a given pain point, conversation learning from question answering, conversation overviews, conversation timelines, conversation anomalies, and other discussion pattern learning techniques. These approaches are referred to as "thick data analytics" since they will find and understand patient insights without losing the context of discussion communities.

Furthermore, Machine Learning (ML) approaches may be utilized as a supplement to learn from recognized data and develop models on identified data. Transfer learning allows these models to be fine-tuned when fresh discussions are incorporated. Experimentation looks at seven insight-driven learning approaches combined with huge geo-located Twitter data to predict the quality of care connected to the present COVID-19 epidemic [11].

The following subsections refer to attempts relating to opioid usage, SNs in hospitals, nature-deficit disorder, and HIV.

Opioid usage

Neo4j is used to collect and handle tweets involving at least one opioid-related term. Opioid (mis)use is a growing public health problem in the United States, hence the purpose of this project is to offer healthcare professionals data on public opioid usage and the geographical distribution of opioid-related tweets. The findings can be used to tailor public health initiatives to communities in high-use regions. Among the results are: (i) During the analysis period, California, particularly the Sacramento area, had the largest number of users (921 people) who sent 2,397 opioid-related tweets. (ii) When compared to the total number of tweets in the state, North Carolina has the greatest proportion (17%) of opioid-related tweets. (iii) The greatest opioid-related user group on Twitter has 42 members, and the most often discussed topic in this group is the negative impacts of Percocet and Tylenol [12].

The opioid pandemic poses a significant public health risk to worldwide communities. Due to limited research and current monitoring limitations, the function of SM in aiding illegal drug trafficking is largely unclear. A computational methodology for automatically detecting illegal drug adverts and vendor networks is described. On a huge dataset gathered from Google+, Flickr, and Tumblr, the SVM and CNN-based approaches for detecting illegal drug adverts, as well as a matrix factorization-based method for locating overlapping groups, have been thoroughly tested. Pilot test results show that these computational approaches may accurately identify illegal drug advertisements and vendor-community. These strategies have the potential to increase scientific understanding of the role SM may serve in prolonging the drug misuse pandemic [13].

Hospital Social Network

The idea of deploying SM in the health sector is now capturing the interest of numerous clinical specialists all over the globe. Solutions for healthcare SNs are evolving to improve patient care and education. However, many clinical operators are hesitant to utilize them since they do not meet their needs, and they are considering developing their own SN platforms to facilitate social science studies. In this situation, one of the primary difficulties is the administration of created Big Data (BD), which has a large number of relations. As a result, typical relational database management solutions are insufficient. The goal of the scientific effort presented in [14] is to demonstrate that NoSQL graph DBMSs may solve such a problem, opening the way for future social science research. The findings of the experiments suggest that Neo4j, one of the leading NoSQL graph DBMS, facilitates the handling of healthcare SN data while also ensuring adequate performance in the context of future social science investigations.

Nature-deficit disorder

The body of evidence supporting the notion that restorative surroundings, green fitness, and nature-based activities improve human health is growing. Nature-deficit disorder, a media phrase used to reflect the negative consequences of people's estrangement from nature, has yet to be legally recognized as a clinical diagnosis. SM, such as Twitter, with its potential to collect BD on public opinion, provides a platform for investigating and disseminating information about the nature-deficit disorder and other nature–health topics. An analysis has been conducted on over 175,000 tweets using SA to determine if they are favorable, neutral, or negative, and then the influence on distribution was mapped.

SA was utilized to analyze the consequences of events in SNs, to examine perceptions about products and services, and comprehend various elements of communication in Web-based communities. A comparison between nature-deficit-disorder "hashtags" and more specific nature hashtags was used to provide recommendations for enhanced distribution of public health information through message phrasing adjustments. Twitter can increase knowledge of the natural environment's influence on human health [15].

HIV

Since the 1990s, the number of new HIV cases and outbreaks have remained steady near 50,000 each year. To increase epidemic containment, public health interventions aimed at reducing HIV spread should be created. Online SNs and their real-time communication capacities are evolving as fresh venues for epidemiological research. Recent research has shown that utilizing Twitter to study HIV epidemiology is feasible. The utilization of publicly accessible data from Twitter as an indication of HIV risk is presented as a novel way of identifying HIV at-risk groups. Existing methodologies are improved by providing a new infrastructure for collecting, classifying, querying, and visualizing data. But also to demonstrate the possibility of classifying HIV at-risk groups in the San Diego area at a finer degree of data resolution [16].

The HIV epidemic is still a major public health issue. According to recent statistics, preventive initiatives do not reach a large number of persons in susceptible communities. To allow evidence-based prevention, researchers are investigating novel ways for identifying HIV at-risk groups, including the use of Twitter tweets as potential markers of HIV risk. A study on SN analysis and machine learning demonstrated the viability of utilizing tweets to monitor HIV-related threats at the demographic, regional, and SN levels. This methodology, though, exposes moral dilemmas in three areas: (i) data collection and analysis, (ii) risk assessment using imprecise probabilistic techniques, and (iii) data-driven intervention. An examination and debate of ethics are offered based on two years of experience with doctors and local HIV populations in San Diego, California. [17].

3.2.6   Influence detection

SN analytics has been a critical issue with extensive material exchange from SM. In SM, the established directed connections govern the flow of information and represent the user's impact. Various issues arise while dealing with data due to the massive amount of data and the unstructured nature of exchanging information. Graph Analytics (GA) appears to be an essential tool for tackling issues such as constructing networks from unstructured data, inferring knowledge from the system, and studying a network's community structure. The suggested method sought to identify Twitter influencers based on follower and retweet connections. Various graph-based algorithms were used to the acquired data to identify influencers and discussion communities in the Twitter [18].

Another work suggested applying a Markov Chain Approach to BD ranking systems. To begin, a transition matrix is constructed, which contains a determined score between each eligible pair of nodes on a large-scale network. The static distribution of the Markov Chain is then computed using the transition matrix. The results of this matrix are ranked to determine the most important users on the network. With certain example datasets, an algorithmic solution on the Neo4j platform was tested. The suggested strategy appears to be promising for mining prominent nodes in large SNs, based on experimental findings [19].

3.2.7   Fake and controversial information

This section elaborates on attempts that deal with fake and controversial information with the next subsections reporting on attempts to detect fake users and news.

Fake users

With the rapid increase in the number of Web users, SN platforms have been among the most important forms of communication all over the world. There are several notable participants in this industry, such as Facebook, Twitter, and YouTube. Most SN platforms include some type of measure that may be used to characterize a user's popularity, like the number of followers on Twitter, likes on Facebook, and so on.

Yet, it has been seen in past years that many users seek to influence their popularity by using false accounts. A certain technique discovers all false followers in a social graph network using attributes related to the centrality of all nodes in the graph and training a classifier using a subset of the data. Employing solely centrality measurements, the proposed approach detected false followers with a high degree of accuracy. The suggested technique is general in nature and may be applied regardless of the SN platform in the hand [20].

The Ethical Journalism Network defines disinformation as deliberately manufactured and publicized material meant to misguide and confuse others. It tries to manipulate the uneducated into believing lies and has a bad societal impact. Fake SM users are seen as famous, and they distribute false information by making it appear genuine. The goal of this research project is to increase the accuracy in detecting fake users [20]. By utilizing several centrality measures supplied by the Neo4j graph database and two additional datasets a classification technique (Random Forest) was incorporated to detect fake users on SM [21].

Several SM platforms that permit content distribution and SN interactions have risen in popularity. Although certain SM platforms enable distinguishing group tags or channel tagging using '@' or '#', there has been a lack of user personalization (a user must recollect a person by their SM name on sign-up). An alternative approach is presented, in which users may store and then seek their friends by using names they use to identify them in real life i.e. nicknames. Furthermore, the suggested approach may be utilized in chatting systems for identifying individuals and tagging friends. Similar to how the '@' is used on SM platforms such as Facebook and Twitter [22].

Pathogenic SM profiles, like terrorist supporter accounts and fake media publishers, have the potential to propagate disinformation virally. Early identification of pathogenic accounts is critical since they are likely to be major sources of spreading illicit information. The causal inference approach was used in conjunction with graph-based metrics to discriminate pathogenic from ordinary users within a short period. Both the supervised and semi-supervised techniques were applied without taking into consideration network data or content. The benefits of the suggested framework were demonstrated using results from a real-world Twitter database. A precision of 0.90 and an F1 score of 0.63 were achieved, posing an improvement of 0.28 compared with previous attempts [23]. The F1 score is an ML statistic that may be applied to classification models. The F1 score is simply the harmonic mean of Precision and Recall and is calculated using Eq. 1 [24].

$$F1 = \text{Score} = 2 \times \frac{\text{Precision} \times \text{Recall}}{\text{Precision} + \text{Recall}} \tag{1}$$

Fake news

Comments and information posted on the internet and other SM platforms impact public opinion about prospective treatments for detecting and healing illnesses. The spread of these is similar to the spread of fake news regarding other vital areas, such as the environment. To validate the suggested technique, SM networks were employed as a testing ground. Twitter users' behavior was examined using an algorithm. A dynamic knowledge graph technique was also incorporated to describe Twitter and other open-source data such as web pages. Furthermore, a real example of how the corresponding graph structure of tweets connected to World Environment Day 2019 was utilized to construct a heuristic analysis. Methodological recommendations showed how this system may enable the automating of operations for the development of an automated algorithm for the identification of fake health news on the web [25].

The analysis of enormous graph data sets has become a significant tool for comprehending and affecting the world. The utilization of graph DBMSs in various use cases such as cancer research exemplifies how analyzing graph-structured data may help find crucial but hidden connections. In this context, an example demonstrated how GA might assist cast light on the functioning of SM troll networks, such as those seen on Twitter.

GA can efficiently assist businesses in uncovering patterns and structures inside linked data. That way it allows them to make more accurate forecasts and make faster choices. This necessitates effective GA that is well-integrated with graph data management. Such an environment is provided by Neo4j. It offers transactional and analytical processing of graph data, as well as data management and analytics capabilities. The Neo4j graph algorithms are the primary ingredients for GA. These algorithms are effectively implemented as parallelized versions of typical graph algorithms, tuned for the Neo4j graph database. The design and integration of Neo4j graph algorithms were discussed and its capabilities were displayed with a Twitter Troll analysis showcasing its performance with a few massive graph tests [26].

3.2.8   Information modeling

Another work provided a smart tool for SM opinion mining. Novel algorithms were utilized for the hybridization of ontological analysis and knowledge engineering approaches. Also, NLP methods were incorporated for extracting the semantic and affective components of semi-structured and unstructured textual data. These techniques increase the efficiency of SM content-specific data processing as well as the fuzziness of natural language. This approach attempted to transform the RDF/OWL-ontology into a graphical knowledge base. Furthermore, the study proposed a method for inferences on the ontology repository. The method is based on transforming Semantic Web Rule Language (SWRL) constructs into Cypher language features [27].

Microblogging is a popular SN medium where users express their thoughts and ideas. As a result, the vast information on microblogging platforms means that social science academics can have a plethora of study material. A framework for interpreting user viewpoints and detecting complicated linkages in the form of knowledge graphs was suggested. The framework's two core objectives are SA and knowledge graph creation to assist in a better analysis of microblogging data.

The Skip-gram model was used to produce the word embedding matrix and the Bi-LSTM model to accomplish stance classification. When compared to Naive Bayes and SnowNLP, Bi-LSTM performed better in categorizing diverse feelings. Relationships between distinct users were retrieved from their microblogs by identifying certain strings, and user attitudes were then merged into the extracted information. Neo4j was used to create a knowledge network of user opinions. Social science researchers may detect patterns in communication and undertake additional data analysis using the knowledge collected by this framework [28].

Online geographic databases have grown in popularity as a vital repository of material for both SNs and safety-critical systems. To obtain and store information from primary available geographical sources, a common and flexible geographic model is provided. Graph databases are useful for geographic-based systems because they can model massive volumes of interrelated data. They also provide improved performance for a variety of spatial queries, such as identifying the shortest path route and executing high-concurrency relationship queries. The current status of geographic information systems was examined, and a single geographic model known as GeoPlace Explorer (GE) was created. GE can import and store symbolic geographical data from a variety of internet sources including both relational and graph databases, in which many stress tests were conducted to determine the benefits and drawbacks of each database system [29].

To keep up with the expanding amount of multimedia content on smartphones and mobile devices, an integrated strategy for semantic indexing and retrieval is necessary. A generic framework was proposed that unifies available image and video analysis tools and algorithms into a unified semantic annotation. Thus, a retrieval model was produced that implements a multimedia feature vector graph representing several layers of media material, structures and characteristics. These feature representations, when combined with Artificial Intelligence (AI) and ML, can produce precise semantic indexing and acquisition. An introduction to the multimedia analysis framework is also presented, as well as a definition of the multimedia feature vector graph framework. A method for rapid indexing and retrieval of network structures was proposed to satisfy specific needs on smartphones and particularly applications in SM. Experiments were presented to demonstrate the method's efficiency, efficacy, and quality [30].

3.2.9   Environmental and disaster management

Environmental monitoring is considered one of the most effective ways to safeguard living spaces from possible hazards. Traditional approaches relied on documenting judgments of observed things on regular ions on specific themes may generate hybrid observation recordings. For instance, observations might be improved by obtaining Twitter messages from specified locations about certain themes, such as marine environmental activities. SA on tweets was used to show popular sentiment on environmental issues. In addition, two hybrid data sets were investigated. To handle this data, a Hadoop cluster was used, and NoSQL and relational databases for storing data scattered among nodes. The potential of SM analytics and distributed computing was used to compare SM public sentiment with scientific data in real time. This demonstrated that citizen science reinforced with real-time analytics can comprise a unique way of monitoring the environment [31].

Several catastrophic events strike the world every year, killing thousands and destroying tens of billions in buildings and infrastructures. Disaster effect reduction is critical in today's societies. The importance of Information and Communication Technology (ICT) in disaster mitigation, intervention, and recovery evolves. There is a plethora of disaster-related data available, including response plans, event records, simulation data, SM data, and Internet sites. Current data management solutions have limited or no integration features. Furthermore, current improvements in cloud applications, BD, and NoSQL open the door to new disaster data management methods.

A Knowledge as a Service (KaaS) architecture for catastrophe cloud data management (Disaster-CDM) was suggested, with the goals of (i) storing massive volumes of multi-source disaster-related data, (ii) enabling search, and (iii) fostering their interoperability and integration. The data is saved in the cloud utilizing a blend of relational and NoSQL databases. A case study was provided demonstrating the usage of Disaster-CDM in a simulated environment [32].

Earthquakes have recently become a very major subject. In SM, earthquake-related news is always displayed first. Building an earthquake knowledge graph can assist in dealing with seismic news. Leveraging news text data to build an earthquake knowledge graph using the Bi-directional Long Short Term Memory-Conditional Random Field (BiLSTM-CRF) model was proposed. The BiLSTM-

CRF model recognized the topics and writes the entities and their types to a table. Then the entities and their links, as well as the structure of the resource description framework, have been merged to create an earthquake knowledge graph on Neo4j [33].

Information has become more accessible and faster than before since more and more people use the web and SM as sources of news. The growth of the internet and SM has also given a voice to a much larger audience. As a result, each user has the opportunity to function as an active correspondent, generating a large amount of data on live occurrences. Twitter data was gathered and analyzed to detect, evaluate, and present instances of social turmoil in three countries, namely India, Pakistan, and Bangladesh [34].

3.2.10 Profiling and criminality

An analysis of the utilization of Twitter to evaluate if the personality traits of high-performing Navy officers may be determined was conducted. The names of high-performing Navy personnel were obtained from publicly available Navy promotion lists, and those identities were then used to explore Twitter. The aim was to locate prospective accounts belonging to these individuals. Data from Twitter accounts that could be recognized as belonging to Navy personnel were then analyzed. Each user's degree of personality characteristics was computed based on the Five Factor Model. The results were maintained in a graph database. It was found that it is feasible to properly compute a user's personality based on textual analysis of their Twitter activity. Yet, it was proven that this approach is insufficient to detect particular qualities for distinguishing individuals that belong to the Navy [35].

The damage inflicted by cybercriminals is overwhelming, and the number of victims is growing at an exponential rate. SM play a significant part in its proliferation by disseminating criminal information. As a result, identifying these cybercriminals has become a top priority. Despite the efforts, there is no efficient and scalable solution. This seems to be due to the enormous number of real-time streams that pass through SM. The need for real-time solutions based on BD discovering applications in Twitter's SN is highlighted by a distributed real-time architecture. Apache Spark and Kafka Ecosystems were used as a solution to this problem. Furthermore, ontology semantics were incorporated to save assertions in a Neo4j. That way the consistency of the claims as well as the inference of additional information was ensured [36].

Analysts today manually search SM networks for conversations about planned cybersecurity threats, perpetrator strategies and tools, and possible victims. Lincoln Laboratory has proven the capacity to automatically find such talks from English posts in Stack Exchange, Reddit, and Twitter, utilizing ML methods. Hackers frequently utilize SM networks to plan cyber assaults, share techniques and tools, and find possible victims for coordinated strikes. Analysts reviewing these exchanges, can convey information about a dangerous action to system administrators and give them a forewarning about the attacker's skills and purpose.

System administrators must prevent, repel, and identify cyber intrusions, as well as adjust in the aftermath of successful attacks. Advanced warnings allow system administrators to focus on certain assault component types, time periods, and targets, allowing them to be more effective. A widespread denial-of-service assault and website defacement can be mitigated by observing SM and private group chats. Even, in case an assault is successful the security officials can briefly halt certain foreign traffic to the inflicted sites. There were teams prepared to react to attacks and repair or recover websites. Monitoring SM networks is a useful way for detecting hostile cyber conversations, but analysts presently lack the necessary automated tools [37].

A visual analytics system called Matisse was presented for exploring worldwide tendencies in textual information streams with special applicability to SM. It was promised to enable real-time situational awareness through the use of various services. Yet, interactive analysis of such semi-structured textual material is difficult due to the high throughput and velocity of data This system provided (i) sophisticated data streaming management, (ii) automatically generated sentiment/emotion analytics, (iii) inferential temporal, geospatial, and term-frequency visualizations, and (iv) an adaptable interaction scheme that enables multiple data views. The evaluation took place with a real use case, sampling 1% of total Twitter recorder data during the week of the Boston Marathon bombings. The suggested system also contained modules for data analytics and associations based on implicit user networks in Neo4j [38].

3.2.11 Attractions and tourism

Preliminary findings from collecting, analysing, and displaying relationships among comment threads on the Smart Tourism Website were provided. The focus was made on user interactions by conveying queries that comprise the user experiences and opinions regarding the attractions of various tourist places. Moreover, the specification of a conceptual data model and the creation of a data processing workflow formed a prototype system. Such a system enables the capture, analysis, and querying of the implicit SN that is determined by the relationships between user comments. Also, specialized software tools were implemented for graph databases and complicated network analysis [39].

The mission of museums is to inspire tourists. The growing use of SM by museums and tourists may open up new avenues for collecting evidence of inspiration. An inquiry into the viability of extracting expressions of inspiration from SM using a system of knowledge patterns from FrameNet was undertaken. Essentially, a lexicon constructed around models of usual experiences took form. Museum employee interpretation of inspiration with algorithmic processing of Twitter data was matched. Balance was accomplished by utilizing prototype tools to update a museum's Information System. New possibilities were examined through the utilization of SM sources which forced museum personnel to focus on the origin of inspirations and its significance in the institution's connections with its tourists. The tools gathered and analyzed Twitter data relating to two incidents: (i) by working with museum specialists, the utility of discovering expressions of inspiration in Tweets was investigated, and (ii) an assessment utilizing annotated material yielded an F-measure of 0.46, showing that SM may be a viable data source [40].

There has been much debate on how museums may provide more value to tourists which can be simply compensated for using instrumental measures such as attendance data. Questionnaires or interviews can give more in-depth information about the influence of museum activities. Yet, they can be unpleasant and time-consuming, and they only offer snapshots of visitor attitudes at certain periods in time. The Epiphany Project studied the viability of applying computational social science approaches to detect evidence of museum inspiration in tourist SM. Inspiration is described in a way that is useful to museum activity, as are the stakeholders that could benefit, their needs, and the vision for the system's design [41].

SM users may post text, create stories, be co-creators of these stories, and engage in group message sharing. Users with a large number of message exchanges have a substantial, selective effect on the information delivered across the SM. Businesses and organizations should consider segmenting SM web pages and linking user opinions with structural indicators. Sections of a web page containing valuable information should be recognized, and an intelligent wrapping system based on clustering and statistics could be suggested to do so automatically. The goal is to gather information on business or organization services/goods based on real-time comments provided by users on SM. Experimentation on Facebook with posts for a hotel booking website named Booking.com took place. Implications include the formation of a web user community with common interests that are related to a product/service. As a result of the comment responses on SM can be utilized in tourism and other activities, leading to trust increase, as well as loyalty and trustworthiness improvement [42].

3.2.12 Metrics and altmetrics

The conventional citation count, peer review, h-index, and journal impact factors are utilized by policymakers, funding agencies, institutions, and government entities to evaluate research findings or effects. These impact measurements, often known as bibliometric indices, are restricted to the academic community and cannot give a comprehensive picture of research influence in the public, government, or corporate sectors. The recognition that scholarly effect extends beyond the scientific and academic domains has given rise to a branch of scientometrics known as alternative metrics, or altmetrics. Furthermore, academics tend to focus on measuring scientific activity using SM, like Twitter. However, since they lack explicit connections to the underlying source, these count-based assessments of influence are susceptible to misinterpretation.

A traditional citation graph is enlarged to a heterogeneous network of publications, scientists, venues, and organizations based on more trustworthy SM sources such as headlines and blog sites. The suggested technique has two aspects. The first is the combination of bibliometric with SM data. The second analyzes how common graph-based metrics might be used to predict academic influence on a heterogeneous graph. The results revealed moderate correlations and favorable relationships between the generated graph-based metrics and academic influence, as well as a reasonable prediction of academic impact [43].

Ranking profile influence is a major topic in SM analysis. Until lately, influence ranking was exclusively based on the structural aspects of the fundamental social graph. As online SM like Reddit, Instagram, and Twitter are recognized largely for their functionality, there has been a noticeable move to the next logical stage where network functionality is considered. Yet, unlike structural rankings, functional rankings are destined to be network-specific because each SM platform provides distinct interaction opportunities.

Finally, seven first-order Twitter influence indicators were examined and a framework was proposed for calculating higher-order equivalents by offering a probabilistic assessment methodology. Simulations with a Twitter subgraph of real-world important accounts showed that a single measure integrating structural and functional properties beats the others [44].

3.3   Results and analysis

This section describes the resultant categorization of Neo4j use cases in SM, as well as briefly discusses the significance of each research effort and its key aspects. Providing a taxonomy raises and supports the necessity of leveraging contemporary technology to improve information modeling and retrieval procedures. Simultaneously, data analytics and the associated data mining processes may result in performance increases due to the numerous advantages of NoSQL databases over relational databases.

According to Table 3.1, there is multiple uses case of Neo4j implementations covering a variety of research topics. For topic detection, an implementation called "Topicbert" functions as a transformer for transfer learning-based memory-graph approach for identifying multimodal streaming SM topics.

Regarding recommendation systems, some implementations utilize association rules mining for defying the most influenced products, an inference mechanism that elevates retail knowledge for recommending goods, and a QGIS integration with Neo4j.

Regarding branding and marketing, a framework ranks web pages calculating the popularity of brands, SA boosts the ability to determine causality between personality-value-attitude utilizing business ads and a model for product feature inference for defining customer preferences. In learning environments, the utilization of SM and Neo4j creates new possibilities for data analysis also considering the formalism of graphs.

Table 3.1. Categorization of recent Neo4j applications in social media

Category	Article citation	Article description
Topic detection and extraction	[3]	TopicBERT: multimodal streaming SM topic detection.
Recommendation systems	[4, 5, 6]	Product-to-product recommendation, Product recommendation based on retail knowledge, Integrated recommendation system using QGIS and Neo4j.
Branding and marketing	[7, 8, 9]	Online marketing brand analysis framework, SA for the causality between personality-value-attitude for business ads, Implicit customer preferences with a feature inference model.
Learning environments	[10]	Social learning through graphs formalism.
Healthcare analytics	[11, 12, 13, 14, 15, 16, 17]	Insight-driven learning with thick data analytics, Combining Twitter and Neo4j for opioid usage in the USA, Detecting illicit drug ads and vendor communities, NoSQL graph database system for a hospital SN, SA related to "nature-deficit disorder", Characterizing local HIV at-risk populations, Evidence-based prevention of HIV risk for vulnerable populations.
Influence detection	[18, 19]	Graph-based analytics for identifying influencers, A Markov chain approach for identifying and ranking influencers.
Fake and controversial information	[20, 21, 22, 23, 25, 26]	Graph centrality measures for fake followers determination, Fake users identification with random forest classifier, Enhancing online interaction by investigating user-defined nicknames as extra identifiers, Determining pathogenic SM accounts, An automated system for fake news detection in healthcare and environment, Elevating comprehension of SM trolls with large graphs.
Information modeling	[27, 28, 29, 30]	Analyzing the semantic content of SM, User opinion analysis through knowledge graphs, Storage and integration of geospatial data in NoSQL, Semantic indexing and content retrieval on smartphone multimedia.
Environmental and disaster management	[31, 32, 33, 34]	Combining hybrid data for environmental monitoring, Disaster data management through knowledge-as-a-service, Social intercourse for earthquake knowledge graph construction, Detecting social unrest areas with NLP.
Profiling and criminality	[35, 36, 37, 38]	Identifying personality characteristics of navy personnel, Profiling potential cyber-criminals, Detecting malicious cyber-discussions, Trend analysis on text streams with multiscale visual analytics.
Attractions and tourism	[39, 40, 41, 42]	SM analytics for smart tourism, Extracting expressions of inspiration regarding museums, Discovering the intrinsic value of museums through SM analytics, Reputation monitoring in tourism services.
Metrics and altmetrics	[43, 44]	Using news and weblogs for assessing scientist social impact, Influence metrics evaluation using Neo4j and Twitter.

Regarding healthcare analytics, four sub-categories were identified related to opioid usage, hospital SNs, nature-deficit disorder, and HIV. Insight-driven learning in thick data analytics focuses on healthcare. Twitter and Neo4j are combined to analyze the use of opioids in the USA. A computational approach searches for illicit drug ads while finding vendor communities. Neo4j utilization enables the analysis of a hospital-based SN. In addition, SA improved data analytics in the nature-deficit disorder theme. Finally, it is important to search and identify local HIV at-risk social atoms while considering the ethical dimensions enabling evidence-based prevention informing vulnerable communities.

Regarding influence detection in SM, graph-based analytics eases the identification of influencers, while another method elevates the use of the Markov chain in BD ranking systems.

In the field of fake and controversial information two categories were created, fake uses and fake news. Graph centrality measures may be used for identifying fake Twitter followers. Also, random forests and other classifiers help distinguish fake users, while user-defined nicknames can act as an extra feature for identifying online interactions. A framework helps identify pathogenic accounts on Twitter, while fake news detection is evaluated in the areas of healthcare and the environment through improved information management. Finally, graphs and Neo4j are utilized to comprehend which text is considered a troll.

Regarding information modeling, software for the semantic analysis of SM content is proposed and graphs constitute an intelligent analysis of user opinion for knowledge extraction. Neo4j is also used for storing and integrating geospatial data in addition to multimedia indexing and semantic context retrieval from smartphones.

Regarding environmental and disaster management, hybrid data is used for monitoring environmental conditions and a framework offers knowledge as a service for disaster-related data. In addition, graphs integrate social intercourse context while enabling the detection of social unrest areas through NLP in SM.

In the field of profiling and criminality detection in SM, a use case for identifying the personality characteristics of Navy personnel is presented, along with a scalable solution for profiling cybercriminals on Twitter. Furthermore, a malicious discussion may be monitored in SM also utilizing possible trends of text streams through visual analytics.

In attractions and tourism, smart tourism is elevated with graphs and SM analytics can also be implemented for museums, extracting expressions of inspiration. Another project discovers the intrinsic value of museums. Monitoring the reputation of tourism services and attractions may involve various benefits for improving their quality.

Finally, metrics and altmetrics may offer valuable insights. Monitoring mainstream news and weblogs may help retrieve indicators for the social impact of scientific research. In addition, Neo4j and graphs enable a high-order approach for the evaluation and definition of influence metrics in SM such as Twitter.

3.4 Conclusion

This chapter highlights the importance of transitioning from SQL to NoSQL databases investigating use cases of Neo4j in the SM domain. A categorization is conceived, based on the context of the state-of-the-art. The intent is to highlight the importance and benefits of NoSQL in contrast to relational databases while presenting its great applicability spread in various modern data management applications.

Therefore, various topics of application have been examined; yet some factors should be also examined when considering the migration to a NoSQL database or even which NoSQL database is more appropriate to use. Indicatively, this section also discusses the comparison of PostgreSQL with Neo4j and MongoDB with Neo4j in two distinct use cases.

To make the most of the large amount of information accessible in today's BD environment, new analytical skills must be developed. SQL databases, such as PostgreSQL, have typically been favoured, with graph databases, such as Neo4j, limited to the analysis of SNs and transportation data. The MIMIC-III patient database is used as a case study for a comparison between PostgreSQL (which uses SQL) and Neo4j (which uses Cypher). While Neo4j takes longer to set up, its queries are less complicated and perform faster compared to PostgreSQL queries. As a result, while PostgreSQL is a solid database, Neo4j should be considered a viable solution for gathering and processing health data [45].

Furthermore, in the smartphone era, geospatial data is vital for building citizen-centric services for long-term societal evolution, such as smart city development, disaster management services, and identifying critical infrastructures such as schools, hospitals, train stations, and banks. People are producing geo-tagged data on numerous SM websites such as Facebook, Twitter, and others, which may be categorized as BD due to the provision of the three key BD attributes: volume, diversity, and velocity. Multiple sources generate vast volumes of heterogeneous data that cannot be classified.

Moreover, data-driven applications demand a longer throughput time. It is quite tough to manage such a big volume of data. Instead of employing a traditional relational database management system, this geotagged data should be maintained using BD management techniques such as NoSQL. As a result, in the context of geographical information systems and all other use cases, it is critical to select the suitable graph database to be used. In [46], for instance, the authors evaluated the performance of MongoDB and Neo4j querying geotagged data to make a more educated selection.

3.4.1 Contribution

This chapter's goal is to expose the reader to graph databases, notably Neo4j and its SM-related applications. It begins by briefly discussing graph databases and Neo4j. It then covers literature on Neo4j use in SM for different applications.

As a result, it categorizes various use cases, resulting in a taxonomy of usages reporting on the findings. Due to the vast diversity of current applications, this study concludes that transitioning to NoSQL systems is important. Since the tendency toward shifting to NoSQL databases, as well as a rising number of research articles that use NoSQL, it is reasonable to infer that NoSQL implementations will almost replace relational databases sooner or later. Healthcare, recommendation systems, environmental/disaster management, and criminality are just a few of the numerous applications employing SM data that use Neo4j. Neo4j exposes new capabilities for better knowledge management and data analytics by being quicker, more scalable, and capable of handling real-world data.

3.4.2 Future work

This study uncovered several issues that require additional investigation. The planned classification of Neo4j apps in SM creates new opportunities to increase the overall number of articles when more are released. In such a scenario, an expanded version of this study might encompass a broader spectrum of recent NoSQL initiatives in the SM domain.

Moreover, after researching Neo4j use cases in the literature, the goal is to develop a measure that takes into account criteria such as: (i) The amount of Neo4j references on websites. (ii) The broad enthusiasm for Neo4j. (iii) The regularity with which technical talks concerning Neo4j are held. (iv) The number of Neo4j job openings provided via SM posts. (v) The number of profiles in professional networks that contain the term Neo4j. (vi) The importance of SNs. In this manner, it is possible to get indicators of its industry applicability penetration. Furthermore, it would allow for more informed judgments when weighing the benefits of switching from SQL to NoSQL, as well as new chances for data analytics enhancement.

References

[1] Robinson, I., J. Webber and E. Eifrem. 2013. Graph Databases. O'Reilly Media. Cambridge, USA.

[2] Guia, J., V.G. Soares and J. Bernardino. 2017. Graph databases: Neo4j analysis. In: ICEIS (1), pp. 351–356.

[3] Asgari-Chenaghlu, M., M.-R. Feizi-Derakhshi, M.-A. Balafar and C. Motamed. 2020. Topicbert: A transformer transfer learning based memory-graph approach for multimodal streaming social media topic detection. arXiv Prepr. arXiv2008.06877.

[4] Sen, S., A. Mehta, R. Ganguli and S. Sen. 2021. Recommendation of influenced products using association rule mining: Neo4j as a case study. SN Comput. Sci., 2(2): 1–17.

[5] Konno, T., R. Huang, T. Ban and C. Huang. 2017. Goods recommendation based on retail knowledge in a Neo4j graph database combined with an inference mechanism implemented in jess. In: 2017 IEEE SmartWorld, Ubiquitous Intelligence & Computing,

Advanced & Trusted Computed, Scalable Computing & Communications, Cloud & Big Data Computing, Internet of People and Smart City Innovation (SmartWorld/ SCALCOM/UIC/ATC/CBDCom/IOP/SCI), pp. 1–8.

[6] Dubey, M.D.R. and S.R.R. Naik. 2019. An Integrated Recommendation System Using Graph Database and QGIS. Int. Research Journal of Engineering and Technology (IRJET), 6(6).

[7] Aggrawal, N., A. Ahluwalia, P. Khurana and A. Arora. 2017. Brand analysis framework for online marketing: Ranking web pages and analyzing popularity of brands on social media. Soc. Netw. Anal. Min., 7(1): 21.

[8] Jang, H.-J., J. Sim, Y. Lee and O. Kwon. 2013. Deep sentiment analysis: Mining the causality between personality-value-attitude for analyzing business ads in social media. Expert Syst. Appl., 40(18): 7492–7503.

[9] Tuarob, S. and C.S. Tucker. 2015. A product feature inference model for mining implicit customer preferences within large scale social media networks. In: International Design Engineering Technical Conferences and Computers and Information in Engineering Conference, 2015, vol. 57052, p. V01BT02A002.

[10] Stanescu, L., V. Dan and M. Brezovan. 2016. Social learning environment based on social learning graphs formalism. In: 2016 20th International Conference on System Theory, Control and Computing (ICSTCC), 2016, pp. 818–823.

[11] Fiaidhi, J. 2020. Envisioning insight-driven learning based on thick data analytics with focus on healthcare. IEEE Access, 8: 114998–115004.

[12] Soni, D., T. Ghanem, B. Gomaa and J. Schommer. 2019. Leveraging Twitter and Neo4j to study the public use of opioids in the USA. In: Proceedings of the 2nd Joint International Workshop on Graph Data Management Experiences & Systems (GRADES) and Network Data Analytics (NDA), 2019, pp. 1–5.

[13] Zhao, F., Skums, P., Zelikovsky, A., Sevigny, L., Swahn, H., Strasser, M., Huang, Y. and Wu, Y. 2020. Computational approaches to detect illicit drug ads and find vendor communities within social media platforms. IEEE/ACM Trans. Comput. Biol. Bioinforma.

[14] Celesti, A., A. Buzachis, A. Galletta, G. Fiumara, M. Fazio and M. Villari. 2018. Analysis of a NoSQL graph DBMS for a hospital social network. In: 2018 IEEE Symposium on Computers and Communications (ISCC), 2018, pp. 1298–1303.

[15] Palomino, M., T. Taylor, A. Göker, J. Isaacs and S. Warber. 2016. The online dissemination of nature–health concepts: Lessons from sentiment analysis of social media relating to 'nature-deficit disorder'. Int. J. Environ. Res. Public Health, 13(1): 142.

[16] Thangarajan, N., N. Green, A. Gupta, S. Little and N. Weibel. 2015. Analyzing social media to characterize local HIV at-risk populations. In: Proceedings of the Conference on Wireless Health, 2015, pp. 1–8.

[17] Weibel, N., P. Desai, L. Saul, A. Gupta and S. Little. 2017. HIV risk on Twitter: The ethical dimension of social media evidence-based prevention for vulnerable populations. In: Proceedings of 50th Int. Conf. on System Sciences, pp. 1775-1784.

[18] Joshi, P. and S. Mohammed. 2020. Identifying social media influencers using graph based analytics. Int. J. of Advanced Research in Big Data Management System, 4(1): 35–44.

[19] El Bacha, R. and T.T. Zin. 2017. A Markov Chain Approach to big data ranking systems. In: 2017 IEEE 6th Global Conference on Consumer Electronics (GCCE), 2017, pp. 1–2.

[20] Mehrotra, A., M. Sarreddy and S. Singh. 2016. Detection of fake Twitter followers using graph centrality measures. In: 2016 2nd International Conference on Contemporary Computing and Informatics (IC3I), 2016, pp. 499–504.

[21] Zhao, Y. 2020. Detecting Fake Users on Social Media with Neo4j and Random Forest Classifier. University of Victoria, Department of Human & Social Development, http://hdl.handle.net/1828/11809

[22] Aggarwal, A. 2016. Enhancing social media experience by usage of user-defined nicknames as additional identifiers for online interaction. J. Comput. Sci. Appl., 4(1): 1–8.

[23] Shaabani, E., A.S. Mobarakeh, H. Alvari and P. Shakarian. 2019. An end-to-end framework to identify pathogenic social media accounts on Twitter. In: 2019 2nd International Conference on Data Intelligence and Security (ICDIS), 2019, pp. 128–135.

[24] Tatbul, N., T.J. Lee, S. Zdonik, M. Alam and J. Gottschlich. 2018. Precision and Recall for Time Series. Mar. 2018, Accessed: Dec. 09, 2021. [Online]. Available: http://arxiv.org/abs/1803.03639.

[25] Lara-Navarra, P., H. Falciani, E.A. Sánchez-Pérez and A. Ferrer-Sapena. 2020. Information management in healthcare and environment: Towards an automatic system for fake news detection. Int. J. Environ. Res. Public Health, 17(3): 1066.

[26] Allen, D., Hodler, E.A., Hunger, M., Knobloch, M., Lyon, W., Needham, M. and Voigt, H. 2019. Understanding trolls with efficient analytics of large graphs in neo4j. BTW 2019.

[27] Filippov, A., V. Moshkin and N. Yarushkina. 2019. Development of a software for the semantic analysis of social media content. In: International Conference on Information Technologies, 2019, pp. 421–432.

[28] Xie, T., Y. Yang, Q. Li, X. Liu and H. Wang. 2019. Knowledge graph construction for intelligent analysis of social networking user opinion. In: International Conference on e-Business Engineering, 2019, pp. 236–247.

[29] Ferreira, D.R.G. 2014. Using Neo4J geospatial data storage and integration. Master's Dissertation. University of Madeira Digital Library, http://hdl.handle.net/10400.13/1034

[30] Wagenpfeil, S., F. Engel, P.M. Kevitt and M. Hemmje. 2021. Ai-based semantic multimedia indexing and retrieval for social media on smartphones. Information, 12(1): 43.

[31] Chen, J., S. Wang and B. Stantic. 2017. Connecting social media data with observed hybrid data for environment monitoring. In: International Symposium on Intelligent and Distributed Computing, 2017, pp. 125–135.

[32] Grolinger, K., M.A.M. Capretz, F. Mezghani and E. Exposito. 2013. Knowledge as a service framework for disaster data management. In: 2013 Workshops on Enabling Technologies: Infrastructure for Collaborative Enterprises, 2013, pp. 313–318.

[33] Sun, X., Qi, L., Sun, H., Li, W., Zhong, C., Huang, Y. and Wang, P. 2020. Earthquake knowledge graph constructing based on social intercourse using BiLSTM-CRF. In: IOP Conference Series: Earth and Environmental Science, 2020, vol. 428(1), p. 12080.

[34] Clark, T. and D. Joshi. 2019. Detecting areas of social unrest through natural language processing on social media. J. Comput. Sci. Coll., 35(4): 68–73.

[35] Ward. 2016. Using social media activity to identify personality characteristics of Navy personnel. Naval Postgraduate School Monterey United States.

[36] Maguerra, S., A. Boulmakoul, L. Karim and H. Badir. 2018. Scalable solution for profiling potential cyber-criminals in Twitter. In: Proceedings of the Big Data & Applications 12th Edition of the Conference on Advances of Decisional Systems. Marrakech, Morocco, 2018, pp. 2–3.

[37] Campbell, J.J.P., A.C. Mensch, G. Zeno, W.M. Campbell, R.P. Lippmann and D.J. Weller-Fahy. 2015. Finding malicious cyber discussions in social media. MIT Lincoln Laboratory Lexington United States.

[38] Steed, C.A., J. Beaver, P.L. Bogen II, M. Drouhard and J. Pyle. 2015. Text stream trend analysis using multiscale visual analytics with applications to social media systems. In: Conf. ACM IUI Workshop on Visual Text Analytics, Atlanta, GA, USA.

[39] Becheru, A., C. Bădică and M. Antonie. 2015. "Towards social data analytics for smart tourism: A network science perspective. In: Workshop on Social Media and the Web of Linked Data, 2015, pp. 35–48.

[40] Gerrard, D., M. Sykora and T. Jackson. 2017. Social media analytics in museums: Extracting expressions of inspiration. Museum Manag. Curatorsh., 32(3): pp. 232–250.

[41] Gerrard, D., T. Jackson and A. O'Brien. 2014. The epiphany project: Discovering the intrinsic value of museums by analysing social media. Museums Web, 2014.

[42] Ntalianis, K., A. Kavoura, P. Tomaras and A. Drigas. 2015. Non-gatekeeping on social media: A reputation monitoring approach and its application in tourism services. J. Tour. Serv., 6(10).

[43] Timilsina, M., W. Khawaja, B. Davis, M. Taylor and C. Hayes. 2017. Social impact assessment of scientist from mainstream news and weblogs. Soc. Netw. Anal. Min., 7(1): 1–15.

[44] Drakopoulos, G., A. Kanavos, P. Mylonas and S. Sioutas. 2017. Defining and evaluating Twitter influence metrics: A higher-order approach in Neo4j. Soc. Netw. Anal. Min., 7(1): 1–14.

[45] Stothers, J.A.M. and A. Nguyen. 2020. Can Neo4j Replace PostgreSQL in Healthcare? AMIA Summits Transl. Sci. Proc., vol. 2020, p. 646.

[46] Sharma, M., V.D. Sharma and M.M. Bundele. 2018. Performance analysis of RDBMS and no SQL databases: PostgreSQL, MongoDB and Neo4j. In: 2018 3rd International Conference and Workshops on Recent Advances and Innovations in Engineering (ICRAIE), 2018, pp. 1–5.

Combining and Working with Multiple Social Networks on a Single Graph

Ahmet Anil Müngen [0000-0002-5691-6507]

OSTIM Technical University, Software Engineering Department, Ankara, Turkey
e-mail: ahmetanil.mungen@ostimteknik.edu.tr

Internet users use social networks for different purposes and with different data. Having the same user accounts in different social networks and combining users' data in a single graph will improve the functioning of recommendation systems and increase the user experience. In this study, the data of thousands of users in nine different social networks were collected and combined in a single graph. Anchors are created between nodes in graphs with different attributes using the previously proposed node alignment and node similarity methods. Node similarity methods have been developed for multiple social networks, and node matching success has been increased. Thus, it has been possible to propose much more successful recommendation systems. In addition, a new alignment method for multiple social networks is proposed in this study. As a result of the study, the success rates of the proposed methods were measured with the actual data collected from social networks. Graphs from multiple social networks have been converted into a single graph. A broad user profile covering more than one social network has been created for users.

4.1 Introduction

Social networks have been one of the most significant technological innovations that entered the internet world in the early 2000s, with the widespread use of the Internet and the increasing penetration rate of mobile devices. In social networks, users can create their pages, visit the pages of their followers, and see their posts. Thus, social networks are channels that allow users to produce content. In this model, the users can publish their material, create friend lists, follow lists, and interact with other users.

People use different social networks according to their needs and purposes. For example, social networks like Instagram, and Flickr, focus on sharing photos. There are also social networks that contain posts limited to a specific character number, called microblogging. Each user can use one or more of the social networks they need as they wish. With the increase in social networks, whose users constantly generate data, the amount of data available on the Internet has increased significantly. Although there are tens of microblogging services globally, Twitter has close to 2 million users, and nearly 500 million tweets are sent every day [31]. One of the largest social networks, Facebook has more than 1.5 billion users, which significantly changed the user's social networking experience. Facebook now has more users than most countries in the world [26]. In Instagram, a Facebook company and another popular photo-sharing network, users have shared over 50 billion photos so far.

Graphs with different features represent social networks because users can share different information using them. Therefore, when comparing the graphs of social networks with each other, it is seen that there are few similarities and almost entirely different graphs.

With the increasing popularity of social networks, mapping users in social networks has recently become an essential issue in academia and the industry. Theoretically, cross-platform discoveries allow a bird's eye view of the user's behavior across all social networks. However, almost all studies that use social network data focus on a few specific social networks [18, 28]. Therefore, using limited data in a small number of social networks causes problems in determining the success of the proposed methods and prevents accurate results. Furthermore, graphs with different features represent social networks because users can share different information using them. Therefore, when comparing the graphs of social networks with each other, it is seen that there are few similarities and almost entirely different graphs.

Identifying and matching the accounts of the same users in different social networks can create a graph with very detailed information for the users, and data mining methods can work more successfully in this new graph. Because users share different information on different social networks, a social network may contain unique data not contained in other social networks. By combining graphs, links and information that have not been obtained before can be obtained, and thus more successful recommendations can be presented to users. The proposed method combines the user's information with information in different social networks in a single graph with a single node.

The paper continues with the following sections. In section 2, node similarity and topological alignment methods are mentioned, and previous studies in the literature are presented. The proposed methods are described in detail in section 3. In section 4, the datasets used in selected social networks and the test results are mentioned. Finally, section 5 is the conclusion part of the study.

4.1.1   Background

Combining different social networks is an increasingly popular topic in recent years. Studies in this area are generally divided into two: Matching the nodes according to their similarity characteristics by looking at their attributes. The other one is alignment to nodes according to connections between the nodes on the graph. Thus, node similarity algorithms have been a frequently studied subject by researchers.

Researchers generally do not need to represent more than one social network because they work with only one social network. Therefore, data representation with multiple graphs is not used much. Each graph representing social networks is independent and often seen as very different. Layers represent different graphs even when social networks are defined by graphs (the same nodes in some graphs may be anchored). A structure with three different graphs and the relationships between graphs are shown in Figure 4.1. Figure 4.1 shows three different social networks. Some users use all three social networks and some users use only two or only one social networks. In the social network at the top, the relationships are through the group. In the middle network, the relationships are bidirectional. In the network at the bottom, the relationships are one-way. In such social networks with different features, accounts belonging to the same user can be matched and connection points can be created.

Proper selection of attributes is vital for most of the studies on social networks. Attributes can be directly selected from the data presented in social networks or obtained as derived. For example, the user's registration date or friends' number is called a direct attribute. The language attribute obtained by

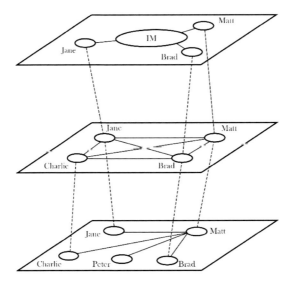

Figure 4.1. Graph representation used in social networks [34].

inferring from the posts is called the derived attribute. It is often preferred to use attributes as a set rather than use them alone and specify rule trees with more than one attribute. While there are similarities between the two nodes, tens of attributes can be used and analyzed. It may not always be possible to match the values in the attributes exactly. Information contained in users' actions or profiles is often tried to find relationships using tagging. Some of the tagging examples are shown in Figure 4.2.

Social network users actively use non-multiple choice data entry fields. Therefore, users can enter the same information in different ways. As a result, it is often not possible to find similarities using classical text-matching methods. Instead, it is necessary to use methods that find partial similarities for almost every field. Tagging users' interests and finding common locations are other problems and cannot be solved successfully by standard methods. The process of detecting and tagging the words (terminological names) frequently used in the user's posts is critical to determine the user's area of interest. Extracting this information with an approach other than labeling is not very possible due to the data structure. Correct labeling is a factor that directly affects the success rate. In Figure 4.2, a demonstration of different users' usage methods of different tags and the relationships between tag cloud and tags is presented.

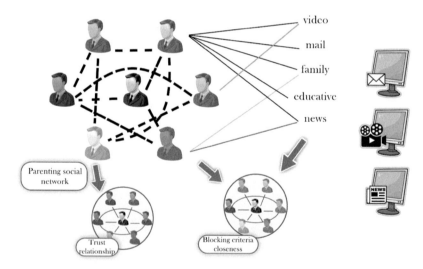

Figure 4.2. Tagging in social networks [12].

4.2   Node similarity methods

It is necessary to compare the attributes of the profiles, especially the area of interest, for the node similarity process. Thus, it can be determined whether two nodes are the same person. Furthermore, the individuals' profiles in different networks can be found at a high rate by analyzing the shared objects (music, movies, activities).

As a result, the combined data can be used for all recommendation algorithms to increase the success rate.

There are many methods to analyze the similarity of nodes in a network. Three of the most popular methods in this area are Adamic Adar, Jaccard, and Common Neighbors methods [23, 32]. Adamic Adar [1] has also been used in many previous studies to predict new connections in the social network and has been reported to have high performance for complex networks. Both the Jaccard Similarity Detection and Adamic Adar were developed to find the similarity between two web pages. Since web pages contain unstructured data like the social network data in the nodes, these methods were also used in this study. Another algorithm used in the study is Friend of Friend (FOAF). Also known as the Common Neighborhood method [3], this algorithm was developed based on the idea that users with many familiar neighbors are more likely to connect in the future. Apart from these, there are other similarity criteria, such as Liben and Kleinberg [6]. Some of these metrics focus on the length of paths between the pair of nodes to predict new connections that will be included in a graph. Nodes can be found in graphs and can be made using calculating the shortest path between nodes [10]. The Restart Random Walk (RWR) algorithm [8] searches for the shortest path by walking randomly on graphs using the Markov chain model. Like RWR, Fouss et al. [13] developed a model to find the most similar node pairs among the nodes in a graph. SimRank [15], on the other hand, accepts the proposition that 'people related to similar people are similar' based on the structural characteristics of the network.

Finally, some methods use other attributes such as messages between users, user ratings, co-authored posts, and common tags besides the graphical structure. Table 4.1 includes the names and some features of the algorithms used for this purpose. Only some algorithms mentioned in this section are included in the table because the table contains only node similarity algorithms. In addition, unlike the existing studies, a vector-based similarity algorithm was also used in this study.

Most of the above methods are used on only one social network link structure. In this study, some of these methods were used but modified for multiple graphs.

Adamic and Adar [1, 2] found four sources of information for a user in their proposed method. The four sources of information that websites mention in their proposed method of finding similarity are links, mailing lists, links, and texts provided by users themselves. Three different data groups were used in this study, namely user-to-user connections, user-related data, and shares. However, since the connections between users are bidirectional, the number of groups can be accepted as four.

N-Gram is frequently used in computational sciences and probability to find n-contiguous sequences in a given text. For written texts, N-gram elements are usually words [7]. Also, web-based unstructured systems generally contain text. Therefore, N-grams are very popular and used for similarity or proximity in these texts.

The task of evaluating the similarity between two nodes in a graph topology is a longstanding problem. It is not possible to quickly find the similarity of data

Table 4.1. Featured node similarity algorithms

Similarity algorithms	Multiple graph	Labeling
SimRank [15]	No	No
P-Rank [39]	No	No
PSimRank [37]	No	No
SimRank++ [4]	No	No
MatchSim [20]	No	No
SimRank* [35]	No	No
RoleSim [16]	No	No
HITS-based Similarity [38]	Yes	No
Random Walk with Restart [8, 19]	No	No
Similarity Flooding [25]	Yes	Labeling
Sampling-based Similarity [40]	No	No
Neighbor Set Similarity [9]	Yes	Labeling
NetSimile [5]	Yes	No
Geometric Feature-based Similarity [14]	Yes	Labeling

based on user behavior with any similarity measure. In addition, applying such analogy methods directly to large graphs brings different problems. Therefore, an effective and scalable link-based similarity model is needed.

SimRank is a general-purpose similarity quantification method that is simple and intuitive and can be applied to graphs. SimRank offers an application that analyzes the similarity of the structural context in which objects occur: object-to-object relationships based on their relationships with other objects. For this reason, SimRank can also be used in areas where partial relationships between objects are required.

For recommendation systems, established known similarities between items and similarities between users may be close to each other. Thus, FOAF is a rational method for describing people, their activities, and their relationships with other people and objects.

FOAF profiles can be used to find all people living in Europe and list the people's friend knows most recognizable among these people. Each profile must have a unique identifier used to identify these relationships. A similarity matrix is used between feature pairs. Similarly, nodes can be found when using machine learning algorithms with the vector space model.

The use of the vector space model to find the same nodes in social networks provides many advantages. Since the vector space model supports many different similarity algorithms, especially cosine, it is also suitable for detecting small changes in values and doing letter-based analysis rather than just getting the intersection set.

4.3   Node alignment methods

Mapping users on a graph in online social networks has recently been popular in academic studies [22]. Figure 4.3 shows an example of the profile-matching method in Liu's paper. Aligning heterogeneous social networks can address data sparsity and can be used to perform meaningful work in social network analysis. For example, profile-based information such as usernames, age information, and demographic information can be matched. Furthermore, many users enter similar data in all profiles on different social networks. Therefore, finding the same users in more than one social network is not an impossible idea [21]. Besides the profile information, the links on the graph also help to find groups [36]. Most of the existing studies do the structural alignment of the networks using matrix multipliers. However, these methods are not easy to scale for large datasets.

Network alignment can be applied locally or globally. For example, LAN alignment aims to align LAN regions precisely [17]. In contrast, most current research [29] has focused on global network cohesion and attempts to make generalized alignments between networks. In general, global network alignment aims to create one-to-one node mapping between two networks (usually worked on one-to-one node mapping, although there are exceptions that produce a many-to-many mapping or align more than two networks).

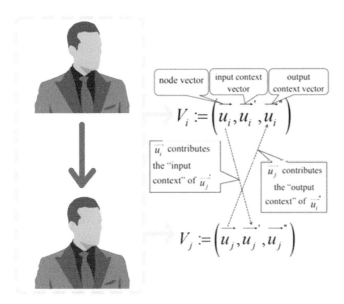

Figure 4.3. Vector-based user profile and links (Liu et al. 2016).

Most of the network alignment methods used in previous studies consist of a node cost function (NCF) and an alignment strategy (AS) [11]. With NCF, methods try to find similarities between nodes in different networks. On the other hand, AS runs the alignment methods based on the analysis performed by the NCF.

Yihan Sun et al. [30] presented an improved NCF in their study. They proposed a method called VOTER for weighted links alignment. Thus, they proposed a novel method for global network alignment with NCFs.

4.3.1   Methodology

In this part of the study, anchor knot methods are first mentioned. Information about the anchor knot method, especially in social networks, is expressed in detail. The section also explains how to use node similarity and topological similarity algorithms. This study proposes an improved method based on the anchor knot method. Although the anchor knot method has been used in social networks before, a method using the anchor method in many more social networks is proposed in this study. While previous studies using the anchor method only focused on node alignment, this study also focused on feature similarity.

4.3.2   Anchor node methods

There are three types of anchor knots used in the study: main anchor, anchor, and candidate anchor. These anchor nodes are the same as each other, and they are just labeled differently. The main anchor is the node that starts as the main in the initial node state and connects the graphs. Candidate anchor is the name given to a similar node which, after assigning the primary anchor node, has two different graphs between the nodes around the primary anchor node and crosses the threshold value when viewed from each attribute direction. If the candidate node exceeds the threshold at the first moment after the transaction, the only node that joins the two graphs is called the anchor node.

The anchor node is the junction point between the graphs for the graphs with more than one social network to converge at the same point. The proposed method must select at least one initial central anchor node for the system to work. Choosing a single or more anchor knots as a starting point does not affect the result. The total number and quality of selected anchor knots are the only factor affecting the result. While choosing the primary anchor node, it is a prerequisite to have an account in all social networks to be taken into account and that these anchor nodes are connected. Although it is unnecessary for all social networks to be connected to a single anchor node, connecting with a single anchor node reduces operational complexity and shortens the processing time. The reason for this is that even though there are intermediate transition points because it is the central point, calculation in the graph is more straightforward than in graphs connected from different places via multiple users.

The data of the first-degree neighbor nodes of the central anchor node are collected. All collected neighbors are called candidate anchor nodes. Candidate anchor nodes are tried to be matched with other candidate anchor nodes in the graph.

The proposed methods aim to align the networks and anchor the same users in different networks, as seen in Figure 4.4. However, each network may not have connections between the same users, and some nodes may be missing or redundant.

An interaction score calculation process has been proposed to calculate scores between nodes. More than ten attributes are used in nine different social networks. The interaction intensity score has been normalized due to social network-based tests and is limited between 1 and 2. However, in topological distance calculations, the length of the paths between the nodes is taken into account. Therefore, in weighted topological calculations, the scores of the routes used between the nodes are also included in the process. In social networks, not every user connected is in the same proximity. Therefore, the weight of the connection is high in cases where the users' relationships are intense in weighted methods. In our study, the weights of the links were measured by the number of common friends between the profiles represented by the two nodes. Also, the number of mutual friends scores normalized to 1 between 2. The formulas used in the interaction density determination method between two nodes are formulas 4.1, 4.2, and 4.3. Formula 4.1 gives the distance as a path between two nodes, formula 4.2 gives the similarity between common attributes, and formula 4.3 gives the interaction score after normalization. Thus, the topological distance with formula 4.1, the similarity score with formula 4.2, the interaction score, and the normalized value of these two were taken.

$$U(x, y) = \sum\nolimits_{\text{All links}} \frac{\text{Used link total point}}{\text{Used link number}} \tag{4.1}$$

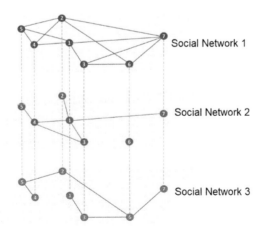

Figure 4.4. Aligned illustration of seven nodes in three social networks.

$$Ep(x, y) = \sum\nolimits_{\text{Interaction ways}} b(x, y) * U(x, y) * \frac{d_T}{s_d} \tag{4.2}$$

$$\textit{Interaction Point} = \frac{Ep(x, y) - Ep(x, y)_{\min}}{Ep(x, y)_{\max} - Ep(x, y)_{\min}} + 1 \tag{4.3}$$

A filter is applied for a small number of people who are thought to distort the average and are expected to be interviewed very frequently. While calculating the average, these people who are frequently interviewed are not included in the calculation. In addition, they cannot be more than ten times the average in order not to disturb the normalization function of 2 nodes with such unusual oversights. Due to the changing usage patterns of social networks over the years, the average calculation method applied for each social network has to be rerun and recalculated at regular intervals.

4.3.3 Attribute similarity-based node matching method

To find the same users in social networks and align the nodes representing the users on graphs, similarities between nodes must be found. There are many node similarity methods proposed for this purpose. However, these methods generally focus on the attributes of the nodes. In this study, a feature vector expressing the properties of nodes is used and developed as a vector similarity method. Thus, we were able to evaluate and measure the vector similarity of many features independently. There is a formula representation of the proposed method in formula 4.4.

$$f_{2lip}(x) = \sum\nolimits_{\textit{Attributes}} b(x) * \frac{U(a,b)}{1 - \dfrac{d_T}{s_d}} \tag{4.4}$$

Table 4.2 lists the algorithms implemented based on attributes.

Table 4.2. Attribute-based method

Attribute	Method
User name	Cosine Similarity Method
User name	N-Gram Similarity Method
Location	Euclidean Location Similarity Method
Popularity	Intersection Method
Using language	SimRank
Active Time Zone	Intersection Method
Interest Tags	Word N-Gram Similarity Method
Personal Statement	TF-IDF

In this study, each attribute was scored with 1 point. Therefore, all the feature similarity methods were arranged to be the lowest 0, and the highest 1. It was determined that it is not appropriate to set a single threshold value for all social networks. The reason for this is that the similarity rate is lower because some social networks provide much more detailed information on some attributes. The most common example in this regard is the city-area location determination from Instagram and Twitter, while the neighborhood-size location can be determined in Meetup and Foursquare. Therefore, the similarity threshold value was determined separately for each social network. In the process of determining each social network separately, separate tests were designed. In these tests, nodes on a low threshold score were manually controlled, and the threshold values per social network were determined by optimizing the F-Measure as maximum. By making the calculation described with all candidate nodes except each node itself, the nodes above the threshold value are matched as anchor nodes, and the node pairs are combined.

4.4   Topological navigation in multiple social networks

The temporal and algorithmic complexity of the topological navigation method increases as the number of social networks counts. In addition, the methods that are suitable and used for dual social networks do not work in multiple social networks due to their nature. For example, in methods where three or more social networks are used, it is impossible to make graph alignments in two dimensions. Therefore, the graphs will have to be curved relative to each other. Moreover, the focused problem cannot be solved by classical methods in horizontal planes, which are generally used.

Graph alignment is done by considering the transitivity rule for a multidimensional graph, and graphs are shown as sections in a three-dimensional representation. In this study, the graphs of social networks were stretched and turned into a sphere, not a horizontal plane. In the proposed model, intertwined spheres with as many as the number of social networks are planned, and since the spheres will be intertwined, it is planned that the node can be brought to the desired location by enlarging the sphere when necessary. As a result of this alignment work, anchor nodes may overlap in more than one graph. Thus, the nodes that are equal to each other will be in line with the graph below when viewed at a 90-degree angle to the sphere. Provided that the anchor nodes overlap, the graph will be aligned by stretching the nodes on the sphere, in other words, by moving them on the graph. Figure 4.5 shows the representation of sample multiple social networks with each other. Every social network has a 3D graph that it can move around. In Figure 4.5, there is an example of a view of the social network.

In Figure 4.6, when looking towards the inner clusters at the arrow level, the profiles of the same person in different social networks will be passed over. In the example figures, representations are made for 4-mesh networks. However, this

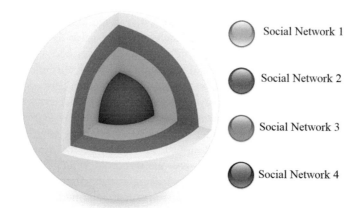

Figure 4.5. Display of spheres on social networks.

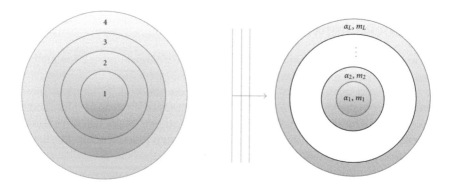

Figure 4.6. Representation of spherical layers [27].

method can be applied to all social networks with more than two. In two social networks, the application of this method will create unnecessary complexity since the 2-dimensional coordinate system is sufficient.

Applying the proposed method to all graphs in all collected data would require a tremendous amount of processing time. The threshold value to be determined is not optimal, and when it is too smaller than it should be, it may lead to an infinite loop. Therefore, the proposed method has to contain many assumptions, filters, and rules. Some of these rules are: The nodes to be aligned can be at most one level away from the primary anchor nodes. Up to 100 nodes can be aligned in a single iteration. Nodes to be aligned are expected to have more than 2x the average interaction amount of all users connected to the anchor node. With these rules, the data set is reduced, and only candidate nodes that can be actual nodes are considered.

4.4.1 Results

In this study, nine different social networks were selected to test the proposed method. This data was collected with a program developed by the authors to conduct the study. This program both collected data with the spider method and used the APIs offered by social networks. The data collected covers the years 2018 to 2020. To apply the proposed method, limited but relevant data was chosen. While selecting the data, the profiles of the researchers and students at the university where the study was conducted and the maximum 2nd level of friends of these profiles were taken into consideration. These social networks are Twitter (T), Instagram (I), Flickr (F), Meetup (M), LinkedIn (L), Pinterest (P), Reddit (R), Foursquare (F), and YouTube (Y). As well as selecting social networks and collecting data, it is very important for the success of the study to determine which attribute to choose in the social network and how the selected attribute is represented in the other social network and to match the attributes. In Table 4.3, the attributes to be used in social networks and the existence of these attributes in other social networks are presented. 'D' means derived, '+' means this data is available, and '-' means this data is not available. 'T' for Text, 'P' for Photo, 'O' for Geo, 'V' for Video, 'OWC' for One Way Connection, 'TWC' for Two Way Connection and, 'G' for Group.

All social networks have usernames that users use to introduce themselves. In Meetup and Foursquare social networks, the success rates of locating and determining the location are very high since they are geo-based social networks. In all networks other than these networks, location determination may not be obtained because it is made from the content. For such cases, a confidence value was applied in the location determination. If there are locations below this value

Table 4.3. Feature-based social networks tables

Feature/SN	T	I	F	M	L	P	R	F	Y
Main Feature	T	P	P	G	T	P	T	O	V
User name (Real)	+	+	+	+	+	+	+	+	-
User name (Register)	+	+	+	+	+	+	+	+	+
Location	+	+	+	+	+	D	D	+	D
Popularity	+	+	D	-	+	-	-	+	+
Language	D	D	D	D	D	D	D	D	D
Active Time Zone	+	+	+	+	+	-	+	-	+
Connection Type	OWC	OWC	OWC	G	TWC	-	-	OWC	-
Interest Tags	D	D	+	+	D	+	D	D	D
Other Social Network Connection	+	-	+	+	-	-	+	-	-
Date of Registration	+	-	+	-	-	-	+	-	-

or two locations that are very close to each other in terms of points but very far geographically, no location determination has been made in the relevant social network for that user. The number makes popularity determination of followers and followers in many social networks. For Meetup, Pinterest, and Reddit, since these numbers are uncertain, all users whose popularity could not be determined were counted as usual. The determination of the languages used was tried to be made by taking the contents shared on the social network and the statements that the user expressed himself if any. Active Time Zone gives the active time zone that the user is using on the social network. The day consisting of 24 hours was divided into six equal parts, and it was tried to determine which part of the user was more active on the social network. Social networks according to interaction types are presented in Table 4.4.

Table 4.4. Interaction type in social networks

Feature/SN	T	I	F	M	L	P	R	F	Y
Like	+	+	+	-	+	+	+		+
Retweet – Repost	+	T	-	-	-	T	-	-	-
Comment	+	+	+	-	+	+	+	-	+
Mention	+	+	+	-	-	+	-	-	+
Favorite	+	-	-	+	-	-	+	-	-

Data from social networks are kept in a relational database, and the data is converted into graphs with dynamically linked lists on an in-memory data grid platform when the calculation is made. After correcting and connecting the graphs, calculations were made. All tests were evaluated on four metrics: Precision (P), Accuracy (A), Recall (R), and F-Measure (F). In formula 4.5, the applied four evolution metrics are given.

$$P = \frac{t_p}{t_p + f_p}, \ R = \frac{t_p}{t_p + f_n}, \ A = \frac{t_p + t_n}{t_p + t_n + t_p + f_n}, \ F_m = \frac{2PR}{P+R} \qquad (4.5)$$

To compare the proposed alignment method, three of the node alignment methods in multiple social networks were selected and tested with the proposed alignment method. In addition, the proposed node similarity method was tested with the link reduction method. The results of the tests performed with the algorithms used for comparison are presented in Table 4.5.

According to Table 4.5, the proposed method obtained the best results, while Liu-Cheung's approach found the second-best results. The worst results belong to the method proposed by Wu-Chien, except for the P value.

An anchor removal test is done by removing a random anchor from the graph after all the nodes in the system are anchored and measuring the system according to finding the removed node pairs. In anchor reduction tests, no properties of the nodes are changed, or the nodes are deleted. Only the anchors are removed.

Table 4.5. Success rates of topological alignment methods

Method name	A	P	R	F
Our Topological Alignment Method	0.79	0.94	0.78	0.85
Liu – Cheung [22]	0.60	0.85	0.62	0.71
Hyrdra [24]	0.60	0.71	0.67	0.69
Wu-Chien [33]	0.60	0.78	0.58	0.67

Although this test is one of the reliable tests in terms of success rate, which nodes will be removed affects the test's success. In this test, the anchor removal process begins with removing anchors in the second or higher degrees of the same node. Thus, the knot-finding process is more successful since the anchor that directly affects the node is not removed. The results of this test with different anchor removal options are shown in Table 4.6.

Table 4.6. Link removing test

Social network pair	SN number	Removed link number	A	P	R	F
Instagram LinkedIn	2	One Link	0.989	0.959	0.950	0.974
Instagram LinkedIn	2	10%	0.989	0.954	0.945	0.971
Instagram LinkedIn	2	20%	0.974	0.964	0.940	0.969
Instagram LinkedIn	2	30%	0.974	0.953	0.930	0.963
Instagram LinkedIn	2	40%	0.974	0.944	0.920	0.958
Foursquare LinkedIn Meetup	3	One link	0.989	0.958	0.950	0.973
Foursquare LinkedIn Meetup	3	10%	0.978	0.946	0.930	0.962
Foursquare LinkedIn Meetup	3	20%	0.972	0.935	0.915	0.953
Foursquare LinkedIn Meetup	3	30%	0.966	0.925	0.900	0.945
Foursquare LinkedIn Meetup	3	40%	0.950	0.919	0.880	0.934

According to Table 4.6, which includes the tests performed with the anchor removed, success rates increase when a few anchors are removed. This is because after all the nodes are matched with a very high success rate, the removed anchor does not disturb the general structure, and the method can easily find the missing anchor.

Limited data collection from two different sides of the world does not give accurate results about the study's success. Although the solution that can be applied to this problem is to collect all the data in the network, it is impossible to collect data with this method due to the data security and API limitations of social networks. For this reason, the data set was selected by applying filters to find those with some similar properties. Thus, care was taken to find datasets close to each other from different social networks. The fact that the data used may positively or negatively affect the success rates.

Although the exact boundaries of confidentiality and usage agreements are not drawn in the data extraction process with the bot, the profiles of volunteer users and open profiles are used in our transactions. While the data of voluntary users were being collected, consent declarations were obtained from the users for these data. In these statements, information is given about the purpose of using the data, how long it will be kept and what will be done after use. All obtained data were anonymized.

4.5 Conclusion

This study proposed unique, inclusive, and prosperous methods for node similarity and alignment. These proposed methods have been transformed into interconnected graphs for social networks with different attributes and different properties and represented by different graphs. In addition, a novel method for node alignment in multiple social networks is proposed. As a result of the study, it has been shown that data from multiple social networks can be combined in a single graph through links. In this way, the applicability of data mining algorithms in multiple social networks has increased. Thanks to this new graph created with data from different social networks, the way to use more successful recommendation algorithms and alternative data mining algorithms have been paved.

The data collected from nine different social networks are used for different purposes and evaluated in different categories, analysis and results such as feature analysis and comparison, which features can be matched, and the compatibility of social networks with other social networks is presented.

One of the side contributions of the study is to provide the opportunity to gather information about the profiles of users with profiles in more than one social network on a single node/graph. The scope of this information provided to the user profile also covers the characteristics of the user's connections in different social networks. Thus, the requesting user can benefit from the services provided by social networks more comprehensively, and problems such as cold start problems can be prevented. The success of the suggested methods has been observed by performing more than 500 tests on different social networks. All tests performed were measured with four criteria frequently used in social networking studies and presented in summary.

When the tests were performed and the proposed method was analyzed, it was revealed that the change in the starting point and the changes in the network affected it. Therefore, the result can be changed according to the starting point. This is a factor that negatively affects confidence in the proposed method. In addition, the proposed method has high computational complexity and needs a tremendous amount of memory than most other similar algorithms.

The number of equivalent studies with which the study can be compared is few. Furthermore, the methods and datasets used by previous researchers and the proposed method in this study are different. For these reasons, a good comparison of the success of the study could not be made.

A more advanced model can be suggested in future studies using other features that are not covered in this study. Also, a real-time method can be proposed using parallel and distributed processing methods and high-performance computing methods. In addition, higher success can be achieved by using a self-feeding neural network for node similarity. The change in the users' usage habits and personal interests over the years is a sociological situation. Therefore, the evolution of social networks can be taken into account, and the addition of the time metric can increase the method's success.

References

[1] Adamic, L. and E. Adar. 2005. How to search a social network. *Social Networks*, 87–203.

[2] Adamic, L.A. and E. Adar. 2003. Friends and neighbors on the Web. *Soc Networks*, 25: 211–230.

[3] Aleman-Meza, B., M. Nagarajan, C. Ramakrishnan, L. Ding, P. Kolari, A.P. Sheth, I.B. Arpinar, A. Joshi, and T. Finin. 2006. Semantic analytics on social networks: Experiences in addressing the problem of conflict of interest detection. *In:* Proceedings of the 15th International Conference on World Wide Web .

[4] Antonellis, I., H. Garcia-Molina and C.C. Chang. 2008. Simrank++: Query rewriting through link analysis of the click graph. Proc. VLDB Endow.

[5] Berlingerio, M., D. Koutra, T. Eliassi-Rad and C. Falousos (n.d.) 2012. NetSimile: A Scalable Approach to Size-Independent Network Similarity. ArXiv abs/1209.2684

[6] Blondel, V.D., A. Gajardo, M. Heymans, P. Senellart and P. Van Dooren. 2004. A measure of similarity between graph vertices: Applications to synonym extraction and web searching. *SIAM Review*, 46(4): 647–666.

[7] Broder, A.Z. 1997. Syntactic clustering of the Web. Comput. Networks.

[8] Cai, B., H. Wang, H. Zheng and H. Wang. 2011. An improved random walk based clustering algorithm for community detection in complex networks. *In:* Conference Proceedings – IEEE International Conference on Systems, Man and Cybernetics, pp. 2162–2167.

[9] Cheng, J., X. Su, H. Yang, L. Li, J. Zhang, S. Zhao and X. Chen. 2019. Neighbor similarity based agglomerative method for community detection in networks. *Complexity*, 2019: 1–16.

[10] Corme, T.H., C.E. Leiserson and R.L. Rivest. 2001. Introduction to Algorithms. First Edition, MIT Press.

[11] Faisal, F.E., H. Zhao and T. Milenkovic. 2015. Global network alignment in the context of aging. Proceedings of the 5th ACM Conference on Bioinformatics, Computational Biology, and Health Informatics, pp. 580–580.

[12] Fernández-Vilas, A., R.P. Díaz-Redondo and S. Servia-Rodríguez. 2015. IPTV parental control: A collaborative model for the Social Web. *Inf Syst Front*, 17: 1161–1176.

[13] Fouss, F., A. Pirotte, J.M. Renders and M. Saerens. 2007. Random-walk computation of similarities between nodes of a graph with application to collaborative recommendation. *IEEE Trans Knowl Data Eng*, 19.3, 355–369.

[14] Ghimire, D. and J. Lee. 2013. Geometric feature-based facial expression recognition in image sequences using multi-class AdaBoost and support vector machines. *Sensors* (Switzerland), 13: 7714–7734.

[15] Jeh, G. and J. Widom. 2002. SimRank: A measure of structural-context similarity. *In:* Proceedings of the ACM SIGKDD International Conference on Knowledge Discovery and Data Mining.

[16] Jin, R., V.E. Lee and H. Hong. 2011. Axiomatic ranking of network role similarity. *In:* Proceedings of the ACM SIGKDD International Conference on Knowledge Discovery and Data Mining.

[17] Kelley, B.P., B. Yuan, F. Lewitter, R. Sharan, B.R. Stockwell and T. Ideker. 2004. PathBLAST: A tool for alignment of protein interaction networks. *Nucleic Acids Res*, 32: 83–88.

[18] Krause, H.V., K. Baum, A. Baumann and H. Krasnova. 2019. Unifying the detrimental and beneficial effects of social network site use on self-esteem: A systematic literature review. https://doi.org/101080/1521326920191656646 24:10–47.

[19] Le, D.H. 2017. Random walk with restart: A powerful network propagation algorithm in Bioinformatics field. *In:* 2017 4th NAFOSTED Conference on Information and Computer Science, NICS 2017 – Proceedings, pp. 242–247. Institute of Electrical and Electronics Engineers Inc.

[20] Lin, Z., M.R. Lyu and I. King. 2009. MatchSim: A novel neighbor-based similarity measure with maximum neighborhood matching. *In:* International Conference on Information and Knowledge Management, Proceedings.

[21] Liu, J., F. Zhang, X. Song, Y.-I. Song, C.-Y. Lin and H.-W. Hon. 2013. What's in a Name? An unsupervised approach to link users across communities. *In:* Proceedings of the Sixth ACM International Conference on Web Search and Data Mining, 495–504.

[22] Liu, L., W.K. Cheung, X. Li and L. Liao. 2016. Aligning users across social networks using network embedding. *In:* IJCAI International Joint Conference on Artificial Intelligence.

[23] Liu, M., B. Lang, Z. Gu and A. Zeeshan. 2017. Measuring similarity of academic articles with semantic profile and joint word embedding. *Tsinghua Sci Technol*, 22: 619–632.

[24] Liu, S., S. Wang, F. Zhu, J. Zhang and R. Krishnan. 2014. HYDRA: Large-scale social identity linkage via heterogeneous behavior modeling. *In:* Proceedings of the 2014 ACM SIGMOD International Conference on Management of Data - SIGMOD '14 .

[25] Melnik, S., H. Garcia-Molina and E. Rahm. 2002. Similarity flooding: A versatile graph matching algorithm. *Data Eng*, 117–128.

[26] Önder, I., U. Gunter and S. Gindl. 2019. Utilizing Facebook Statistics in Tourism Demand Modeling and Destination Marketing. 59: 195–208. https://doi.org/101177/0047287519835969

[27] Saltsberger, S., I. Steinberg and I. Gannot. 2012. Multilayer mie scattering model for investigation of intracellular structural changes in the nucleolus and cytoplasm. *Int J Opt*, 1–9.

[28] Shelke, S. and V. Attar. 2019. Source detection of rumor in social network – A review. *Online Soc Networks Media*, 9: 30–42.

[29] Singh, R., J. Xu and B. Berger. 2007. Pairwise global alignment of protein interaction networks by matching neighborhood topology. *In:* Lecture Notes in Computer Science (Including Subseries Lecture Notes in Artificial Intelligence and Lecture Notes in Bioinformatics).

[30] Sun, Y., J. Crawford, J. Tang and T.M. Milenković. 2014. Simultaneous optimization of both node and edge conservation in network alignment via WAVE. *In:* Algorithms in Bioinformatics: 15th International Workshop, WABI 2014, 16–39. Springer Berlin Heidelberg.

[31] Twitter – Statistics & Facts | Statista [WWW Document] (n.d.). URL https://www. statista.com/topics/737/twitter/#topicHeader__wrapper (accessed 3.3.22).

[32] Wang, J. and Y. Dong. 2020. Measurement of text similarity: A survey. *Inf 2020*, 11: 421.

[33] Wu, S.H., H.H. Chien, K.H. Lin and P.S. Yu. 2014. Learning the consistent behavior of common users for target node prediction across social networks. *In:* 31st International Conference on Machine Learning, ICML 2014.

[34] Yang, S.J.H., J. Zhang and I.Y.L. Chen. 2007. Web 2.0 services for identifying communities of practice through social networks. *In:* Proceedings – 2007 IEEE International Conference on Services Computing, SCC 2007, pp. 130–137.

[35] Yu, W., X. Lin, W. Zhang, J. Pei and J.A. McCann. 2019. SimRank*: Effective and scalable pairwise similarity search based on graph topology. *VLDB J*, 28: 401–426.

[36] Zhang, J. and P.S. Yu. 2015. Integrated anchor and social link predictions across social networks. *In:* IJCAI International Joint Conference on Artificial Intelligence.

[37] Zhang, M., H. Hu, Z. He, L. Gao and L. Sun. 2015. A comprehensive structural-based similarity measure in directed graphs. *Neurocomputing*, 167: 147–157.

[38] Zhang, X., H. Yu, C. Zhang and X. Liu. 2008. An Improved Weighted HITS Algorithm Based on Similarity and Popularity, pp. 477–480. Institute of Electrical and Electronics Engineers (IEEE).

[39] Zhao, P., J. Han and Y. Sun. 2009. P-Rank: A comprehensive structural similarity measure over information networks. *In:* International Conference on Information and Knowledge Management, Proceedings.

[40] Zhou, Y., Y. Deng, J. Xie and L.T. Yang. 2018. EPAS: A sampling based similarity identification algorithm for the cloud. *IEEE Trans Cloud Comput*, 6: 720–733.

Child Influencers on YouTube: From Collection to Overlapping Community Detection

Maximilian Kissgen[1*], Joachim Allgaier[2] and +Ralf Klamma[1]

[1] RWTH Aachen University

[2] Fulda University of Applied Sciences, Human Technology Center,
  RWTH Aachen University

[*] email: maximilian.kissgen@rwthaachen.de

Influencers are gaining a growing impact on the social life and economic decisions of the web population. A special group are child influencers, carrying responsibility and popularity unusual for their young age. However, little research data exists regarding child influencers and their follower communities. Furthermore, social networks continue to restrict independent data collection, making the collection of data sets for analysis difficult. This chapter presents an exemplary approach starting from collection of a larger amount of data from YouTube with respect to the access restrictions. By creation of automatic scripts, targeting child influencers and communities on YouTube, followed by storage on the graph database ArangoDB. Through analyzing the data with overlapping community detection algorithms and centralities for a better understanding of the social and economic impact of child influencers in different cultures. With the authors open source WebOCD framework, community detection revealed that each family channels and single child influencer channels form big communities, while there is a divide between the two. The network collected during the research contains 72,577 channels and 2,025,879 edges with 388 confirmed child influencers. The collection scripts, the software and data set in the database are available freely for further use in education and research.

5.1 Introduction

As social networks continue to grow, the impact of influencers on peoples lives increases. Influencers are people with a strong presence within social networks like YouTube, Instagram or Twitter who are using their reputation, measured in followers, to advertise for products and lifestyles [4]. Marketing campaigns made use of influencers recently, which raised a lot of interest in algorithms identifying actual and potential influencers in social network data sets as well as in measures for the impact of influencers on other persons in social networks [10, 22]. Still, socio-economic research and analysis and advances in mathematical analysis and algorithmic support of influence maximization [5, 28] is only starting to grow together.

With it, non-adult or child influencers also gain more importance, making them an interesting target for economic analysis, but also for social debates. Their young age raises concerns within the public. From the perspective of marketing, child influencers may reach markets already lost for advertisements in traditional media like print and television. From the perspective of their parents, they are generating additional income from a family up to making a fortune. Still, laws for child protection and against child labor are still in place.

However, examination of such child influencers has rarely been done so far as little recent social network data exists in this to base it on. One reason for this is maybe the fact that the phenomenon is quite new and evidence is known mostly on an anecdotal basis. Moreover, the phenomenon of child influencers is spread over a quite high number of social networks making it difficult to collect and analyze data in all of them. Also, challenges can be attributed to social networks restricting access to their application programming interfaces (APIs) and data scraping, making collections more difficult to build. Communities of child influencers provide a means to deduct the actual impact range as well as unifying interests for the audience members themselves. For the understanding of child influencer communities, it is also important to identify overlaps between different child influencer communities, e.g. for revealing hidden marketing activities or to identify the work of social bots [24, 29]. With these results, influencers are first identified through influencer identification and then overlapping community detection algorithms. Algorithms applied in this chapter are Clizz, SLPA and LEMON [12, 13, 26]. Graph databases are a good companion for social media analytics, in general. They provide direct storage of graph data ideally suited for social networks with a lot of augmentation tools for later analysis and visualization. Graph databases like Neo4J and ArangoDB offer a set of social media analytics tools like centrality measures and (overlapping) community detection algorithms. In principle, a limited set of analysis tasks can be already performed in the graph database itself, but there is still a larger need for further support. This makes is necessary to work with external tools in the moment. Yet, there are also some challenges in using graph databases for the research. Many graph databases provide import and export tools for relational data format, e.g. CSV, as well as standard formats for graph data, e.g. JSON. But overall support for researchers is still limited. Writing collection scripts is cumbersome and not well supported in many graph databases. In this chapter, an exemplary way to extract graph data from the social network YouTube through means of its API is shown. Because of the abundance of data, preliminary filtering is applied to receive more child influencer channels. The data is stored on an ArangoDB database to support its graph-like nature and allow for easier access. For the analysis, it made use of the authors overlapping community detection algorithm framework on the Web WebOCD. This chapter provides the following contributions.

- A discussion of social network analysis tools in graph databases with a special focus on (overlapping) community detection algorithms for influencer analysis. The special domain of child influencers is introduced and related work is presented.
- A practical study of collecting, storing, analyzing, and visualizing child influencer data from YouTube with special respect to the graph database ArangoDB, which serves as an example for modern multi-model databases.
- A detailed analysis of the child influencer data set collected from YouTube, mostly performed with the authors open source overlapping community detection algorithm framework for the Web WebOCD.

The rest of the chapter is organized as follows. In the next chapter, the state- of-the-art in graph databases for social network analysis and child influencer research are described as well as the overlapping community detection algorithms used in the chapter. This is followed by Section 1.3 on the used methodology of collecting, storing, analyzing and visualizing the data set. After this, the results are presented and discussed, while the chapter concludes with a summary and an outlook on further work.

5.2 Background

5.2.1 Graph databases

Even though SQL databases still enjoy a lot of use, specialized NoSQL graph databases have already taken a noticeable share in popularity[*], see Figure 5.1. By the time of this publication, they are furthermore still continuing to gain importance[†] as shown in Figure 5.2. In the case of graph databases and social networks, this can be attributed to the graph-like qualities of such networks: often, an account can be represented by a node with additional information linked to other accounts through certain relationships, e.g. being friends with someone. Using graph databases can therefore come with less adaption required towards the structure of the social network. They can also come with performance advantages, e.g. the graph database neo4j was shown to support memory up to the petabyte level and furthermore optimized information search [9], which may be highly relevant for larger networks. An important argument for researchers is that networks like Facebook and YouTube offer APIs with graph-like access already and that graph databases support often needed graph traversal as well a set of algorithms for graph analysis. Through their NoSQL-characteristic, object structures can also change, which may become necessary if a network for example introduces new attributes (e.g. birthdays) or removes certain ones from public availability. Like most SQL databases, graph databases also provide web interfaces for users.

---

[*] https://db-engines.com/en/ranking_trend
[†] https://db-engines.com/en/ranking_categories

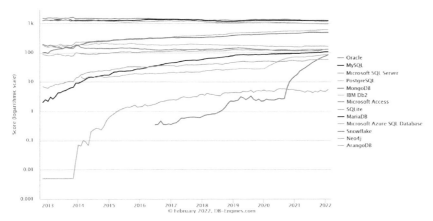

Figure 5.1. A ranking of overall popularity for various SQL and NoSQL databases, including ArangoDB and Neo4j, from 2013 to February 2022 taken from the db-engines website. It is implied that graph databases have reached a noticeable share of popularity with MySQL databases however still being the most popular.

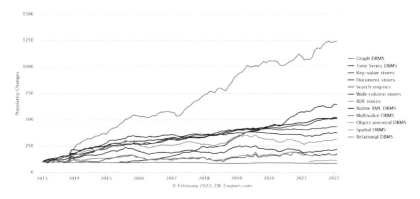

Figure 5.2. A ranking of popularity growth in popularity for SQL and different NoSQL database formats from 2013 to February 2022 taken from the db-engines website. The ranking indicates a strong rise in popularity for graph databases.

In the following, graph analysis algorithms offered by the graph databases Neo4j and ArangoDB will be given a more detailed description: Neo4j provides a set of centrality measures such as PageRank [3], betweenness and degree centralities and community detection in the form of the Louvain method [2], furthermore label propagation and local clustering coefficients algorithms. ArangoDB also offers PageRank, between and furthermore closeness centrality as well as label propagation and SLPA [26] for community detection. While these algorithms can provide helpful initial insights and can be very optimized for parallel processing due to per-vertex computation through architectures like Pregel [16] added by [8] in ArangoDB (although they previously offered them through just AQL functions [15]). It is argued in this approach, with many graph databases offering at most a similar amount of analysis algorithms to those presented, that none of them provide enough different or current ones to perform exhaustive analyses, making the use of additional tools necessary.

5.2.2 Child influencer research in social sciences

A certain number of studies already exist on child influencers on YouTube but this research does not analyze child influencers and their communities together: the majority of the existing studies are detailed analyses on the content and leading persons of a smaller set of child influencer channels, out of which the most influencers examined are in a study by [14] with 20 channels. Their channels content is examined towards brand presence and the influencers themselves concerning their behavior, as well as their motivations for the online presence [17, 23]. Findings from these studies include that parents are often highly involved with the channel, as well as being driving factors for their child's online presence. Advertising and brand presence furthermore seem to play a leading role in the channels content, with some studies noting that the influencers are hesitant to disclose those relationships or do in parts not disclose them at all [6].

Overall, existing research is exploratory and does not concern with greater numbers of child influencers as well as their audiences, leaving the need for more to be done.

5.2.3 Influencer research in computer science

Determining influential nodes and detecting communities within graphs are both important and well-researched topics in computer science. To see whether a node can be considered an influencer in the context of a graph, for example a social network one, various methods for Influencer Detection and Influence maximization exist. The first tends to involve centrality measures that behave in a deterministic fashion, such as coreness or simply taking a nodes degree, while the latter deals with simulating processes such as information diffusion to determine influence [20].

The detection of communities, in either overlapping fashion or not, also exists in a variety of different approaches such as seed set expansion [13] or label propagation [26]. It has also often been applied in the social network context, notably on data sets from Twitter and Facebook, yet as well in Biology and other fields. Influencers gathered from detection or influence maximization can mark initial seed members here, especially for the seed set expansion field, as they will likely dictate a bigger amount of nodes for a community. A current problem for such algorithms is the lack of new social network data sets of sufficient size: Social networks have continuously closed off their APIs or limited access to them. Most of them limit the daily amount of requests made and Facebook and thus also Instagram only allow access to even the public information of a users profile with explicit permission of the user, making collection in greater numbers difficult. Only few bigger networks such as Twitter and YouTube provide researchers greater freedom in using their apis, even though they also have quotas on requests made. Data Scraping is furthermore not considered completely legal in larger quantities and explicitly prohibited in the Facebook and TikTok terms of service[‡,§,¶]. Concentration on Twitter and YouTube may therefore be done more often, with newer larger data sets of YouTube however being from 2012 [27], showing a need for more current ones.

5.2.4 Community detection tools

While many (overlapping) community detection algorithms have an existing implementation available on platforms such as GitHub, and bigger collections of sources such as awesome-community-detection[**] certainly provide easier access to them, there are few applications or libraries that provide a large enough set of different algorithms out of the box, let alone additional functionality.

A popular tool for people from social sciences is SocNetV[††], which offers many needed graph analysis, as well as basic web crawling and graph creation possibilities with a graphical user interface. Included community detection algorithms are however few with a Bron–Kerbosch and an own Triad Census implementation. For Cytoscape there also is a plugin for community detection[‡‡] that features a set of 5 algorithms, such as OSLOM, thus adding to existing features and the UI from the actual application. Furthermore, there are bigger libraries, such as NetworKit[§§] and communities[¶¶] that provide some visual capabilities and a set of algorithms, with communities having 5 different ones and NetworKit having one label propagation and one Louvain method implementation. They however do not have graphical user interfaces. In addition, the mentioned tools often do not provide *overlapping* community detection. The RESTful WebOCD service, developed by [19] at the authors chair aims to fill this niche by providing many overlapping community detection algorithms, and similarly to SocNetV additional graph analysis capabilities, such as many centrality measures for determining influencers. This while also having a graphical user interface in form of a web client with visualization of the results to lessen the need for own programming. Crawling tools are however not included. A remaining problem for all mentioned solutions is that even though many applications provide some newer form of the Louvain or Leiden [21] algorithm and WebOCD now has some recent Local Spectral Clustering [13] and Ant Colony Optimization [18] implementations, the overwhelming number of included algorithms is comparatively old.

Overall, while some as SocNetV and WebOCD try to account for it, few tools feature the means to do at least bigger parts of social network research all with one tool, making it potentially more difficult and time-consuming to do that research without some form of experience

---

[‡] https://www.facebook.com/apps/site_scraping_tos_terms.php
[§] https://www.tiktok.com/legal/tik-tok-developer-terms-of-service?lang=en
[¶] https://www.facebook.com/help/instagram/581066165581870/
[**] https://github.com/benedekrozemberczki/awesome-community-detection
[††] https://socnetv.org
[‡‡] https://github.com/cytoscape/cy-community-detection
[§§] https://networkit.github.io
[¶¶] https://github.com/shobrook/communities

5.2.5   Used community detection algorithms

SLPA

The first algorithm used is Speaker Label Listening Propagation (SLPA) [26]. In the algorithm, every node i will start with a label $l(i)$ unique to the node and each node will contain a memory of size $M_i > 1$ for storing labels. The main loop for each node i is the following: i is marked as a listener to which each neighboring node j, considered a speaker, sends a randomly selected one of its stored labels, resulting in set L_j. The listener will then store the label l_{max} which is the one most often received. If multiple labels share this maximum occurrence, a random one is selected. Each iteration, this is done for every node until a specified k iterations have happened. The final step is then to only use labels with an occurrence probability smaller than a specified $\varepsilon \in [0, 1]$.

The remaining t labels mark the containing communities with affiliation $\dfrac{occc(l(j))}{t}$

for each label $l(j)$ with the number of occurrences $occ(l(j))$. In the worst-case SLPA will take $O(Tn)$ for sparse networks and $O(Tm)$ for others with T being the memory size of a node. SLPA is relatively fast which makes it a good use case for larger graphs, it is however not considered to produce highly accurate communities [25].

CliZZ

The CliZZ algorithm [12] works by locally expanding communities through which it is also able to work on larger graphs. It works by identifying influential nodes and forming communities around them. For this, the algorithm defines its own measure called leadership:

$$f(i) = \sum_{j=1, d_{min}(i,j) \leq \left\lfloor \frac{3\delta}{\sqrt{2}} \right\rfloor} e^{-\frac{d_{min}(i,j)}{\delta}}$$

Here, δ is a chosen constant and $d_{min}(i, j)$ representing the shortest distance between two nodes, where i is the leader candidate. Leadership thus infers how many nodes can be reached by as small a distance as possible from a candidate, with a maximum distance smaller than $\left\lfloor \dfrac{3\delta}{\sqrt{2}} \right\rfloor$. For a leader-node i it then holds that $f(i) > f(j)$ with all other nodes j that the candidate can reach. For every non-leader node i, a vector x^i signifying community affiliation with l entries is initialized with randomly generated normalized entries. From there, the membership matrix is calculated through a random walk like process:

$$x_l^i(t+1) = \frac{1}{\deg(i)+1}\left(x_l^i(t) + \sum_j a_{i,j} x_l^j(t)\right); t \in [0, k]$$

with $a_{i,j}$ being an entry in the graphs adjacency matrix. This is repeated for a specified k times, or when the infinity norm of the difference between an updated membership matrix and the previous one is smaller than a specified precision factor p.

LEMON

The final algorithm makes use of spectral clustering and is called LEMON [13]. It was chosen to detect communities centered around one or a few specific influencers without looking at the whole graph. A spectra defines an l-dimensional subspace in which the closeness of the nodes for the community sub-graph is captured. Initial seed sets must be provided by own means.

Starting from the seed set, a sub-graph S is defined as its ground truth community and iteratively expanded. The procedure initially takes a specified k steps of random walk from the nodes of the seed set resulting in probability vector p_0. From there, a local spectra is generated in multiple steps.

Initially, A_s as a derivation from the normalized adjacency matrix A_s of the sub-graph for the community is calculated:

$$\overline{A}_s = D_s^{-\frac{1}{2}} * (A_s + I) * D_s^{-\frac{1}{2}}$$

Where D_s being the diagonal degree matrix of the sub-graph and D^{-1} signifying taking the square root of all entries of D_s and then taking the inverse. Finally, I is the identity.

The second step is to compute a Krylov matrix $K_{l+1}(A, p_0)$ and, i.e. a basis of l normalized eigenvectors orthogonal to each other.

The third step is a k-loop: Deriving an orthonormal basis of eigenvectors of the matrix yields the first spectra $V_{0, l}$. By multiplying $\overline{A}_s$ with the spectra and taking its orthonormal basis again, the next spectra $V_{1, l}$ is created and so on.

The result is the local spectra $V_{k, l}$, consisting of l random-walk probability vectors. With it, the following linear problem is solved:

$$\min \|y\|_1$$

$$s.t.\ y = V_{k,l}\, x,$$

$$v \geq 0,$$

$$y(S) \geq 1$$

The solution-vector y is a sparse vector in which each element signifies the likelihood of the corresponding node to be in the target community. From there, the global conductance for this iteration is calculated and stored. If this global conductance increases after a local minimum for the first time, the algorithm stops as it has found a reasonably accurate community, repeating it otherwise. According to the authors, the algorithm does not scale with graph size but with community size. This can lead to it being a lot faster than others in application.

5.3   Methodology

For the approach, an ArangoDB database was first set up to store the collected information. Collection itself was handled through a NodeJS-script periodically collecting channels and links, while collecting less information for channels below a 5,000 subscriber threshold and filtering for potential child influencer candidates through regular expressions, relaying collected to the database in the process. A channels subscriptions/favorites/comments led the script to other channels. With help of WebOCD, centrality calculations were run to verify the possible child influencers from the script, followed by overlapping community detection through CliZZ, SLPA and LEMON.

5.3.1   Storage

Method description

As YouTube's data can be considered graph-like with the channels being nodes linked by subscriptions, comments, etc., ArangoDB was chosen, in addition to previous affiliation with the chair, with a single-node setup. This was considered sufficient for the collected amount. The structure of the objects in the database is largely oriented on how objects are structured in the YouTube Data API, which already returns JSON objects. Although the database is schema-less, sticking to a specific structure can be beneficial for analysis and speed up traversal as ArangoDB generates implicit schemas from its documents. For each channel and link a collection was created. To allow for graph representation, traversal and export, the latter collections were labeled as edge collections.

The channel document is given in Figure 5.3, it can either contain just the *statistics* and *contentDetails* part, or additionally (if *subscriberCount* is greater than 5,000) channel *snippet*, *status* and *topicDetails*. The *statistics* contain the amounts of comments, views, videos and furthermore the amount of subscribers. The *contentDetails* contain a list of the channels public playlists ids and the snippet contains the channel title, description and its creation date. In the *status* snippet there is finally information about the channels special properties. The edge documents are represented in Figure 5.4. Comments contain their text, subscriptions their publishing date and favorites the id of the favorited video. The direction of an edge is understood here as from the channel the action was done to towards the one that carried the action out.

Limitations

While sticking closely to the object format of the YouTube Data API arguably makes collection and storage easier, objects may become more verbose than needed. For instance, keeping separate parts, e.g. snippet, in every larger channel object may become problematic for bigger data sets and leave attributes in the less refined returned state may make understanding them more difficult. Finally, having two separate channel objects does not only mean less available information for all channels below 5,000 subscribers, it also requires accounting for two different types of objects when using the information as a channels description is not guaranteed to exist.

```
Channel
{
  "snippet": {
    "title": String,
    "description": String,
    "publishedAt": String,       //Date like '2015-03-20T02:50:24Z'
    "country": String            //Countrycode like 'US'
  },
  "contentDetails": {
    "relatedPlaylists": {        //All elements contain their playlists id
      "likes": String,
      "favorites": String
    }
  },
  "statistics": {
    "viewCount": Int,
    "commentCount": Int,
    "subscriberCount": Int,
    "hiddenSubscriberCount": Bool,
    "videoCount": Int
  },
  "topicDetails": {
    "topicCategories": String[] //List of links to wikipedia topics
  },
  "status": {
    "privacyStatus": String,     //Either 'public', 'private' or 'unlisted'
    "isLinked": Bool,            //Linked to a Google/Google+ account or not
    "madeForKids" : Bool         //YT Kids rules apply, NOT set by owner
  }
}
```

Figure 5.3. The channel object containing snippet-, content-, statistics-, topic- and status-information. The basic structure is similar to the channel object in the YouTube API, and a channel with below 5,000 subscribers will just hold *statistics* and *contentDetails.*

```
     Comment                  Favorite                 Subscription
{                        {                          {
  "videoID": String,       "videoID": String,         "publishedAt": Date
  "value": String          "videoTitle": String,    }
}                          "videoTags": String[]
                         }
```

Figure 5.4. The comment, favorite and subscription objects: While the last just holds its publishing date, thefirst two are related to a video and hold the corresponding id as an attribute. Favorites additionally hold the videostags, while comment objects contain the comment text.

5.3.2 Collection

Method description

YouTube offers an API for developers to access specific information. As most networks it has introduced (daily) quotas for requests made to the API, making fetching of friends or connected groups of people difficult. Likes were left out of collection as they are set to be private by default and thus likely to be few. Collection itself was realized via NodeJS. While the full source code is available on github[***], a pseudo

---

[***] https://github.com/MaxKissgen/ytdapi_fetch

code representation is given in Figure 5.5. NodeJS was chosen due to the availability of libraries supporting client-sever communication. Furthermore, ArangoDB as

```
1   async function scheduler(seedUserIds) {
2       channelQueue = seedUserIds;
3       while (channelQueue.length !== 0) {
4           if(channelExistsAlready(channelQueue.front()) === false) {
5               waitUntilNextDayIfQuotaExceeded();
6               channel = CollectChannel(channelQueue.front());
7               saveChannel(channel);
8           } else {
9               channel = getSavedChannel(channelQueue.front());
10          }
11          if(channel.isInfluencer === true) {
12              if(isUnlikelyChild(channel)) {
13                  unlikelyChildQueue.enqueue(channel);
14                  continue;
15              } else if(CommentsDisabled(channel) === false) {
16                  waitUntilNextDayIfQuotaExceeded();
17                  comments = collectComments(channel);
18                  saveComments(comments);
19                  channelQueue.enqueue(comments.channels);
20              }
21          } else if(favoritesAvailable(channel) === false) {
22              waitUntilNextDayIfQuotaExceeded();
23              favorites = collectFavorites(channel);
24              saveFavorites(favorites);
25              channelQueue.enqueue(favorites.channels);
26          }
27          if(subscriptionsAvailable(channel) === false) {
28              waitUntilNextDayIfQuotaExceeded();
29              subscriptions = collectSubscriptions(channel);
30              saveSubscriptions(subscriptions);
31              channelQueue.enqueue(subscriptions.channels);
32          }
33          channelQueue.dequeue();
34
35          if (channelQueue.isEmpty() && !unlikelyChildQueue.isEmpty()) {
36              channelQueue.enqueue(unlikelyChildQueue.front());
37              unlikelyChildQueue.dequeue();
38          }
39      }
40  }
```

Figure 5.5. The scheduler from the script in pseudo code: The top channel from the queue is chosen, startingwith a set of seed channel ids. From there a reg-ex check is made whether the channel can be an influencer and thena potential child influencer. Only then, additional channel information is collected, along with Youtube comments.Otherwise, favorites are collected and the channel moved to the *unlikelyChildQueue* if it is a potential influencer. Subscriptions are fetched for every type of channel.

well as YouTube offer libraries for the language. Due to the quota set by Google, collectible information per day had a set limit and a high collection speed, as it could be done lower-level, was not an objective as the script would idle most of the time. The script maintains two queues with channel ids, the *channelQueue* is an initial location for newly found channels and contains seed channels at the start while the *unlikelyChildQueue* is a backup of channels over 5,000 subscribers that are unlikely to be child influencers.

The main loop starts with the scheduler taking a channel id from the channelQueue, or, if empty, from the *unlikelyChildQueue*, fetching its statistics part.

If the *subscriberCount* is greater than 5,000, the channel is likely an influencer and the snippet is fetched. Whit it, two reg-ex checks are then made to first see if it is a music channel or an official presence of a personality from another platform/medium and then whether it is a potential child influencer. If the first is true, the channel gets discarded as such channels were found to cultivate big audiences during testing and may act as a sink. Otherwise, the favorites of the channel are fetched and it is moved into the *unlikelyChildQueue*. If the second is true, comments for that channel are fetched alongside its subscriptions and it is then deleted from the queue. Only English and German child influencers were considered in this approach, the phrases and words were gathered from an initial sample and reg-ex's check for mentions of parents and immediate family, of favorite activities such as toys and pretend play and the own age as an important point. These expressions do only represent the sample and not child influencer context as a whole, achieving the latter was considered out of scope.

Channels below 5,000 subscribers only retain the statistics part and are deleted from the queue afterwards. For any of the comments, subscriptions and favorites, not all are fetched, as loading any additional page (of 50 objects) increases the quota usage. The script is set up to consider additional mentions of channels in the queue and saves the id of the last retrieved page for a channel. For each iteration, up to 250 comments, 100 subscriptions and only the first 50 favorites (as getting favorites is costly due to having to get the related video first) of channels are fetched. Channels and the different edge objects are sent to the connected ArangoDB database after processing, leaving their JSON structure mostly intact. At any point during the collection, the collection quota set by YouTube can run out and the scheduler will idle until the next day pacific time when the quota is reset.

Limitations

The script implements snowball sampling [7] in that collected objects deliver the next to be examined, and thus likely much more than one other channel. This results in a certain bias towards specific types of content, however allowing to potentially get more child influencers, and in an exponentially in-creasing sample size, making the script unable to run for an indefinite amount of time. The regular expressions also work toward that bias and even though bigger, non child influencer channels are potential sinks they still provide important and connecting parts of a network that are ignored with this method. Furthermore, the fetching of only some links for some channels and never all of a certain type can skew the actual strength of bonds between them and others. This and the option for users to make their subscriptions etc. private/unavailable may ignore some connections completely.

5.3.3 Analysis

Method description

From importing the resulting data of the query as an XGMML file and uploading it to WebOCD, a selection of centrality calculations are run to determine whether possible child influencers from the collection could be considered as such. WebOCD was chosen because of the greater variety of overlapping community detection algorithms compared to other tools as well as its affiliation with the chair. The centralities include closeness, coreness, neighborhood coreness, degree, eccentricity, eigenvector and betweenness centralities and a combination of them through Borda count. From the remaining child influencer nodes, actual child influencers are filtered out manually according to the following principle:

- The channel is not an online presence of (broadcast) networks
- The channel has to feature children/non-adults in at least 70% of its videos OR
- It has done so in the past up to or since a certain point in time
- The same children/non-adults appear regularly in content from the channel

After determining the actual child influencers, on a smaller subgraph with all child influencers and considerably less non-influencer nodes CliZZ, SLPA and LEMON are used to detect communities around them.

Limitations

While the selection criteria can provide a good way to define what channel can be considered a child influencer and which is not, it is debatable if they can cope with all possible channels, possibly leaving out or wrongly including child influencers with irregular but prominent children appearances, and cannot give an objective definition. Implications of the community detection results on a small graph will likely be less for the bigger graph, let alone for the whole of YouTube. A greater variation of used algorithms, also non-overlapping ones might furthermore deliver more refined results.

5.4   Results

Upon termination of the collection script, 72,577 channels have been collected. Furthermore 2,025,879 edges, with 1,176,617 subscriptions, 750,688 comments and 98,574 favorites. Out of the collected channels, 12,627 were considered possible influencers, therefore having over 5,000 subscriptions, while the rest of the objects then had reduced information due to them being below. The complete results can be found on GitHub[†††].

5.4.1   Influencers

Discussion

The collection script found 1,335 channels for the English and 302 channels for the German regular expression. Most of the English child influencer channels disclose their location as the United States of America, further notable countries include Great Britain, the Philippines Australia. Others only occur below three times. The German child influencer channels, with the exception of one Austrian and one Swiss channel, were all German.

The centrality detection results showed the probable child influencer channels to be scoring high in centrality rank, for the top channels by Borda count of the centralities

---

[†††] https://github.com/MaxKissgen/UCISNC_thesis_results

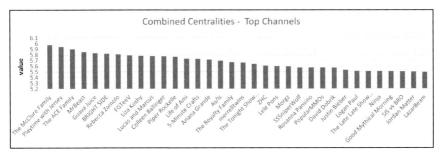

Figure 5.6. The most influential channels by Borda count value.

refer to Figure 5.6. In addition, early added channels ranked rather high in comparison to others. Under the top channels, many are entertainment oriented, e.g. MrBeast and Guava Juice. However, especially under the top 5 channels, there are family channels or such that feature children such as The McClure Family, which were added early in collection.

Out of the channels, each above 5,000 subscribers, child influencers were then selected according to the criteria mentioned in Section 5.3. After this further selection, 388 channels remained which will be considered as child influencers in this approach. For the English regular expression 347 channels remain, and 41 for the German regular expression. Although most of the channels are indeed speaking the corresponding language, some are included that speak a different language, yet make use English or German wording respectively.

Comparing the found possible child influencers with the number of those after manual filtering, the pre-selection through regular expressions can be considered somewhat fitting, however the number difference is large enough to suggest that expressions can be improved upon in terms of accuracy, while the smaller number of found German channels overall can additionally hint at more seed channels being needed and a more verbose regular expression. That script selected them as being possible influencers overall can be seen as accurate with the centrality results.

Limitations

The smaller number of child influencers possibly makes them less representative of the child influencer spectrum on YouTube as a whole. This sentiment can be extended to the number of found possible child influencer channels from the script. That there are just English and German speaking channels, makes the results likely to leave out important findings as parts of the world speaking different languages or highly differing dialects are not considered. Even though the centrality rank results support them to be influencers, the apparent higher ranking of early added channels may have distorted results in their favor and with the used centralities, channels with very strong influence through their content on however few other users are likely less favored than those just having a lot of connections and paths to others.

5.4.2 Community detection

Discussion

For community detection, a sub graph of the completed one was used, containing all 388 child influencer nodes and 8,045 nodes below 5,000 subscribers. Each SLPA, CliZZ were run on it and LEMON was run with the leading nodes of the resulting communities as seeds.

SLPA found 5,084 communities, most were isolated nodes, and was run with a probability threshold of 0.09 and set the node memory size of 5,000. The cover can be seen in Figure 5.7. Here, leading nodes are considered those with the highest affiliation and most direct followers. The biggest community with 878 nodes that of the McClure Family, which was one of the starting nodes for collection, the community was in large parts made up of family vlog channels. Different family members also have their own channels and they and their communities shared this big one. Playtime with Jersey was however the exception in that it had his own community but shared members with those of the McClure one. The next big community is the one lead by Piper Rockelle with 244 members, including mostly channels led by singular children that do do challenges and sometimes music videos. Their followers were included as well. Other family channels and their followers, lead by The Ohana Adventure have their own community with 162 members. No large differences in content were found compared to the McClure community. Many German child influencer channels, lead by Mamiseelen also had a community with 46 members. Exceptions to this were Family Fun with 28 members and TeamTapia with 42 members. The members consisted mostly of direct followers of the channels. Most other more isolated nodes with few connections to the rest with a comparatively large amount of followers, e.g. The Ace Family, were mostly affiliated to an own community along with those followers. CliZZ found 5,180 communities in the subgraph, most being isolated nodes. Refer to Figure 5.7 for a visual overview. The influence factor parameter δ was set to 20, the iteration boundary for the membership calculation to 1,000 and the membership precision factor was set to 0.001. Overall, community affiliation for many communities for many nodes was very low, often 0.01 and less. This made for less isolated and thus more overlapping communities than in the SLPA results. For Conciseness' sake, these affiliations will not be discussed more closely, concentrating instead on those with larger values, considered here to have *significant* affiliation. The most notable community "leaders" were the same as those in the previous results. However this time with mostly less members of *significant* affiliation in general. The McClure Family was again a leader of a community, consisting of family channels and their direct followers, this time with 488 and therefore less members of significant affiliation. In general, a set of smaller family channels now group together without the bigger ones, while those were mostly leaders of their own communities. Piper Rockelle had a bigger amount of 306 significant members, compared to the previous results. The channels community again influencers (and their direct followers) of similar content, with less of these influencers being a part of the community this time. For the German influencers, the results were mostly similar to SLPA, however succumbing to the trend of less big communities with

Figure 5.7. The biggest connected component of the sub graph with SLPA results on the left and Clizz on the right: The Piper Rockelle community is portrayed in red/dark-blue top left, consisting of other children doing pranks/challenges. The McClure Family community is in light-blue/red on bottom right with family channels. The German Mamiseelen community has most German channels and is on the upper left corner in green/dark-green.

Mamiseelen now having a community of 25 significant members for example, many of them still being German child influencer channels.

Selecting the leaders from the previous algorithms' results as seeds, along with some of the nodes with the highest affiliation, the LEMON Algorithm was run. To allow for better comparison between different affiliations, the logarithm to base of 10 was taken of its and the results then normalized. Subspace dimension and the number of random walk steps were each chosen to be 3 and the initial probabilities took a nodes degree into account, the maximal expansion step of the seed set was set to 6 per iteration. Significant members of communities are again those with over 0.01 affiliation. LEMON was run with minimal community size of 1,300 and maximal size of 5,000 for the McClure family. The results correspond with those of SLPA, including most family vlog channels and their direct followers, having 1,447 significant members. Many direct followers of single family members are again part of the community but direct followers from only Playtime with Jersey were not significant members this time, even though the influencer channel was. For Piper Rockelle the algorithm was run with minimum community size of 1,000 and maximum size of 5,000. Along similar members compared to the previous results, as high amount of family channels was now featured as well, raising the number to 1611 significant members. Child influencer channels with a single child as the content producer with challenges and in parts singing however had higher affiliation values. In addition, under the included family channels there were more concentrating on challenges or skits in comparison to the McClure community which featured mostly family vlog channels. For the German child influencer community run for Mamiseelen, the algorithm was run with minimum size of 10 and maximum size of 2,100. The community featured this time include most German influencer channels with 111 significant members. With lower affiliation, Family Fun and Team Tapia were included as well.

The results suggest that communities form, even though bigger influencers (in terms of direct followers) can still form their own or impact communities, around certain topic types of influencer channels such as family-vlog ones more than just around big channels. In consequence, it is implied that users that consume content of a child influencer channel centered around a specific topic are likely to consume content or are related to users that consume similar content from other child influencer channels as well. Other findings are that there is a divide of interest between people that follow family channels and those that follow singular child influencers, such as Piper Rockelle for English channels, while this seems to not be the case for German channels.

Limitations

Overall, implications of the community detection on YouTube as a whole are, as mentioned in Section 5.3, limited by the small number of community members and nodes compared to the complete graph. This is especially the case for the smaller communities as the lack of members places high importance on the ones that do exist. This possibly resulted in isolated nodes that are not isolated overall and less accurate communities. With regards to the collection process, it has to further be said that the concentration on just child influencer channels may hide links between their followers or themselves through other types of channels, such as sport or music themed ones, alongside with the quota-reserving measures taken in Section 5.3, possibly distorting the resulting community structures.

5.5 Conclusion

Social network research and graph databases are indeed good companions. It was demonstrated in this chapter that one can make use of a graph database like ArangoDB to collect, store, analyze and visualize a data set from a social network like YouTube. However for it, a lot of efforts have to be made like scripting the collection scripts and utilizing an overlapping community detection algorithm framework as part of a not yet integrated tool chain. One can argue that graph databases should focus on storing graph data but, similar to relational databases, tighter integration of processing tools should however be at least discussed.

Also demonstrated was the use case of child influencer community detection as a means to interdisciplinary research between socio-economic research and computer science in the field of emerging computational social science [1, 11]. The work may be ground laying and exemplary for further collaborative research in this area. Although, the use case from a more practical and technical perspective was described and discussed, the impact on research tools usable for non-computer scientists is already obvious. Managing data collection with parameterized scripts allows research to address new collections with regular expressions for search terms. Storing data in managed databases frees researchers from handling versions of excel sheets and plain data files. Computational tools configured and executed through a user friendly Web interface ease the work of researchers from installing and managing tools on their computers. They can share and discuss results openly with colleagues world wide, just by exchanging URLs if they like.

Last but not least some interesting results on child influencer communities were computed and visualized, indicating that communities form around child influencer channels with similar topics and that there is a potential divide between family channel communities and singular child influencer channel communities. Child influencer are still an emerging phenomenon and as the numbers show, with likely more in the US than in Europe. One can clearly identify influencers in child influencer communities but with lower numbers of members results can get inaccurate.

This leads to a number of challenges in using graph databases as tools for social network analysis. Automation or integration is the key for many labor-intensive and complicated steps in the research procedures. The more automation is possible, the more researchers will be able to use computational tools based on graph databases. Sharing and discussing results demands more platforms for storing and exchanging results among researchers. Open data and data infrastructures on the national and international level are key. Ideally, social network providers would provide better support in both research automation as well as result sharing. Since this is not very likely to happen, researchers must find ways to overcome current lack of support. Open source for graph databases and the supporting infrastructure is a must in the future. ArangoDB is open source and has a lively community. WebOCD is still a research and teaching tool with a much more limited number of developers, but hopefully more researchers and developers are supporting WebOCD or similar platforms.

References

[1] Alvarez, R.M. (editor). 2016. Computational Social Science: Discovery and Prediction. -Analytical Methods For Social Research. Cambridge University Press, Cambridge, United Kingdom and New York, NY, USA and Port Melbourne, Australia and Delhi, India and Singapore, first published. edition.

[2] Blondel, V.D., J.-L., Guillaume, R. Lambiotte, and E. Lefebvre. 2008. Fast unfolding of communities in large networks. *J. Stat. Mech.*, 10: 10008.

[3] Brin, S. and L. Page. 1998. The anatomy of a largescale hypertextual web search engine. *In:* International World Wide Web Conference Committee, editor, Computer Networks and ISDN Systems, pp. 107–117.

[4] Brown, D. and N. Hayes. 2008. Influencer Marketing. Routledge.

[5] Chen,W., Y. Wang and S. Yang. 2009. Efficient influence maximization in social networks. *In:* Elder, J. (Ed.). Proceedings of the 15th ACM SIGKDD International Conference on Knowledge Discovery and Data Mining. p. 199, New York, NY. ACM.

[6] de Veirman, M., S. de Jans, E. van den Abeele and L. Hudders. 2020. Unravelling the power of social media influencers: A qualitative study on teenage influencers as commercial content creators on social media. *In:* Goanta, C. and Ranchordás, S. (Eds.). The Regulation of Social Media Influencers. pp. 126–166. Edward Elgar Publishing.

[7] Goodman, L.A. 1961. Snowball sampling. *The Annals of Mathematical Statistics*, 32(1): 148–170.

[8] Grätzer, S. 2017. Implementing Pregel for a Multi Model Database. Master thesis, RWTH Aachen University.

[9] Guia, J., V. Goncalves Soares and J. Bernardino. 2017. Graph databases: Neo4j analysis. *In:* Proceedings of the 19th International Conference on Enterprise Information Systems, pp. 351–356. SCITEPRESS – Science and Technology Publications.

[10] Kiss, C. and M. Bichler. 2008. Identification of influencers – Measuring influence in customer networks. *Decision Support Systems*, 46(1): 233–253.

[11] Lazer, D., A. Pentland, L.A. Adamic, S. Aral, A.-L. Barabási, D. Brewer, N.A. Christakis, N.S. Contractor, J. Fowler, M. Gutmann, T. Jebara, G. King, M. Macy, D. Roy and M. van Alstyne. 2009. Social science: Computational Social Science. *Science*, 323(5915): 721–723. New York, N.Y.

[12] Li, H.-J., J. Zhang, Z.-P. Liu, L. Chen and X.-S. Zhang. 2012. Identifying overlapping communities in social networks using multiscale local information expansion. *The European Physical Journal B*, 85(6).

[13] Li, Y., K. He, K. Kloster, D. Bindel and J. Hopcroft. 2018. Local spectral clustering for overlapping community detection. *ACM Transactions on Knowledge Discovery from Data*, 12(2): 1–27.

[14] López-Villafranca, P. and S. Olmedo-Salar. 2019. Menores en youtube, ¿ocio o negocio? an´alisis de casos en españa y eua. *El Profesional de la Información*, 28(5).

[15] Dohmen, L. 2012. Algorithms for Large Networks in the NoSQL Database ArangoDB. Bachelor's thesis, RWTH Aachen, Aachen.

[16] Malewicz, G., M.H. Austern, A.J.C. Bik, J.C. Dehnert, I. Horn, N. Leiser and G. Czajkowski. 2010. Pregel: A system for large-scale graph processing. *In:* Elmagarmid, A.K. and Agrawal, D. (Eds.). Proceedings of the ACM SIGMOD International Conference on Management of Data (SIGMOD 2010), SIGMOD '10, pp. 135–146.

[17] Gorshkova, N., L. Robaina-Calderín, and J.D. Martín-Santana. 2020. Native advertising: Ethical aspects of kid influencers on youtube: Proceedings of the ethicomp* 2020. *In:* ETHICOMP* 2020, pp. 169–171.

[18] Shahabi Sani, N., M. Manthouri and F. Farivar. 2020. A multi-objective ant colony optimization algorithm for community detection in complex networks. *Journal of Ambient Intelligence and Humanized Computing*, 11(1): 5–21.

[19] Shahriari, M., S. Krott and R. Klamma. 2015. Webocd. *In:* Proceedings of the 15th International Conference on Knowledge Technologies and Data-driven Business, pp. 1–4. ACM.

[20] Srinivasan, B.V., N. Anandhavelu, A. Dalal, M. Yenugula, P. Srikanthan and A. Layek. 2014. Topic-based targeted influence maximization. *In:* 2014 Sixth International Conference on Communication Systems and Networks (COMSNETS), pp. 1–6. IEEE.

[21] Traag, V.A., L. Waltman and N.J. van Eck. 2019. From louvain to leiden: Guaranteeing well-connected communities. *Scientific Reports*, 9(1): 5233.

[22] Tsugawa, S. and K. Kimura. 2018. Identifying influencers from sampled social networks. *Physica A: Statistical Mechanics and its Applications*, 507: 294–303.

[23] Tur-Viñes, V., P. Núñez-Gómez and M.J. González-Río. 2018. Kid influencers on youtube: A space for responsibility. *Revista Latina de Comunicación Social*, (73): 1211–1230.

[24] Wagner, C., S. Mitter, C. Körner and M. Strohmaier. 2012. When social bots attack: Modeling susceptibility of users in online social networks. *In:* Proceedings of the 2nd Workshop on Making Sense of Microposts held in conjunction with the 21st World Wide Web Conference 2012, pp. 41–48.

[25] Xie, J., S. Kelley and B.K. Szymanski. 2013. Overlapping community detection in networks. *ACM Computing Surveys (CSUR)*, 45(4).

[26] Xie, J., B.K. Szymanski and X. Liu. 2011. Slpa: Uncovering overlapping communities in social networks via a speaker-listener interaction dynamic process. *In:* 2011 IEEE International Conference on Data Mining Workshops (ICDMW).

[27] Yang, J. and J. Leskovec. 2012. Community affiliation graph model for overlapping network community detection. *In:* Ding, Y. (Ed.). Proceedings of the ACM SIGKDD Workshop on Mining Data Semantics, pp. 1170–1175, New York, NY.

[28] Yang, W., J. Ma, Y. Li, R. Yan, J. Yuan, W. Wu and D. Li. 2019. Marginal gains to maximize content spread in social networks. *IEEE Transactions on Computational Social Systems*, 6(3): 479–490.

[29] Yuan, X., R.J. Schuchard and A.T. Crooks. 2019. Examining emergent communities and social bots within the polarized online vaccination debate in twitter. *Social Media + Society*, 5(3): 205630511986546

Managing Smart City Linked Data with Graph Databases: An Integrative Literature Review

Anestis Kousis [0000-0002-1887-5134]

The Data Mining and Analytics Research Group, School of Science and Technology, International Hellenic University, GR-570 01 Thermi, Thessaloniki, Greece
e-mail: a.kousis@ihu.edu.gr

The smart city paradigm is about a better life for citizens based on advancements in information and communication technologies, like the Internet of Things, social media, and big data mining. A smart city is a linked system with a high degree of complexity that produces a vast amount of data, carrying a very large number of connections. Graph databases yield new opportunities for the efficient organization and management of such complex networks. They introduce advantages in performance, flexibility, and agility in contrast to traditional, relational databases. Managing smart city-linked data with graph databases is a fast-emerging and dynamic topic, highlighting the need for an integrative literature review. This work aims to review, critique, and synthesize research attempts integrating the concepts of smart cities, social media, and graph databases. The insights gained through a detailed and critical review of the related work show that graph databases are suitable for all layers of smart city applications. These relate to social systems including people, commerce, culture, and policies, posing as user-generated content (discussions and topics) in social media. Graph databases are an efficient tool for managing the high density and interconnectivity that characterizes smart cities.

6.1 Introduction

Cities are critical for our future, as most people live there. They are complex systems that have been constructed through a long process of many small actions

[1]. Thus, the city operation is challenging, and new methods emerge to manage and gain insights from massive amounts of generated data. The smart city concept aims to make inhabitants' lives more sustainable, friendly, green, and secure. It lies in the explosive growth of Information and Communication Technologies (ICT) due to the advancement of revolutionary technologies like the Internet of Things (IoT), social media, and big data mining which are considered the e-bricks for smart city development. Citizens post opinions on social media and participate in online discussions generating new opportunities for improving government services while boosting information distribution and reachability. A smart city is a social and technical structure, carrying a very large number of connections. Social media may contain unstructured data regarding any smart city application layer. For example, users may discuss optimal resource management issues related to water or electricity supply.

Graph theory emerges as the default way of organizing the complex networks that city services need and possibly the only realistic way of capturing all this density and interconnectivity [2]. When dealing with linked data, one compelling reason to choose a graph database over a relational database is the sheer performance boost. Another advantage of graph databases is that they allow schemas to emerge in tandem with the growing understanding of the domain, as they are naturally additive. This means that they are flexible in adding new kinds of nodes and relationships to an existing structure, without operating problems on existing queries. Furthermore, the schema-free nature of the graph data model empowers evolving an application in a controlled manner. Thus, in smart city data, graph databases are ideal for modeling, storing, and querying hierarchies, information, and linkages.

Smart city data management with graph databases is an emerging topic that would benefit from a holistic conceptualization and synthesis of the literature. The topic is relatively new and has not yet undergone a comprehensive review of the literature. Each source of urban data varies in scale, speed, quality, format, and, most importantly, semantics and reflects a specific aspect of the city [3]. Knowledge graphs uphold linked city data regardless of type, schema, or any other traditional concern [4]. What is fascinating about smart city applications is that all these graphs are interconnected. For example, the devices and users may be represented on a map, depicting the device, citizen, and location graphs in a single view. Many independent smart city applications can be found by reviewing the related literature, but there is an absence of holistic, synthetic, and integrated proposals. This work aims to synthesize new knowledge and probe future practices on the topic by weaving together ideas from the literature into a unique integrated model.

The rest of the article is structured as follows: The second section lays out a conceptual framework for smart cities, social media, knowledge graphs, and graph database research. The third section presents the materials and methods used in the current study. The fourth section outlines an overview of the related literature in the field. The next section composes the emerging ideas found in the literature into an integrative framework for managing smart city-linked data with graph databases. Finally, in the last section, conclusions and future study directions are discussed.

6.2   Background

6.2.1   Smart cities

As Townsend states: "cities accelerate time by compressing space, and let us do more with less of both" [1]. As cities grow, the need to change the way they operate becomes more and more vital. Never before has there been such an opportunity to do it [5]. In recent years, the question of how ICT might be used to improve a city's performance and promote efficiency, effectiveness, and competition has arisen [6]. The smart city concept was proposed by International Business Machines Corporation (IBM) in 2008 [7] as the potential solution to the challenges posed by urbanization [8].

D'Aniello et al. [9] consider the smart city as an evolving, dynamic system with a dual purpose: supporting decision-making and enriching city domain knowledge. The smart city strives to offer citizens a promising quality of life by merging technology, urban infrastructures, and services to dramatically enhance function efficiency and answer residents' requests via resource optimization [10]. The smart city operation is based on six pillars: Smart People, Smart Economy, Smart Mobility, Smart Living, Smart Governance, and Smart Environment [11]–[13] as illustrated in Figure 6.1. Li et al. [14] argue that only by considering all

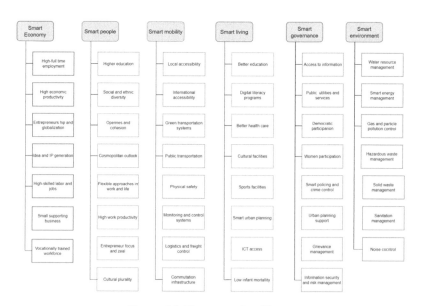

Figure 6.1. The smart city pillars.

of these factors equally can a smart city accomplish social fairness, economic development, and sustainability.

The smart city is based on a 3D geospatial architecture that allows for real-time sensing, measurement, and data transmission of stationary and moving objects [14]. Large amounts of data generated can be converted into valuable information that will guide decision-making through proper management and processing. D' Aniello et al. [9] represent the smart city as an adapted system that operates in three phases, as shown in Figure 6.2. The ubiquitous sensor network (hard sensing) and social media interactions capture real-time data in the first phase (soft sensing). These data produce streams, which should be processed in the second step to change them into valuable information that may be used to make decisions. Knowledge leads to actions in the city during the third phase.

The overall architecture of a smart city is based on the three-layered hierarchical model of the IoT paradigm as illustrated in Figure 6.3. Townsend [1] states that these three layers allow renovating of governments by design, transforming the way they work internally and together with citizens and other stakeholders. The three layers of the smart city structure are:

1. The "instrumentation" layer: Data from the urban environment and citizen social interaction are collected in this layer. Smart devices with distributed data collection are in charge of gathering and storing data from the monitored region locally. Images, video, music, temperature, humidity,

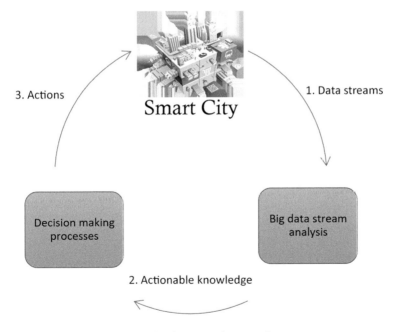

Figure 6.2. The smart city operation.

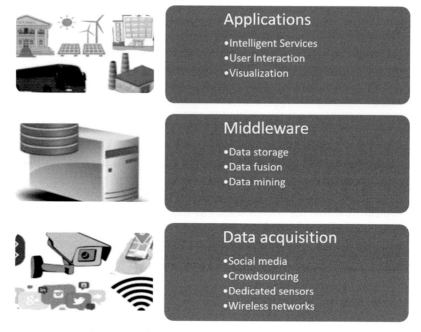

Figure 6.3. The three layers of smart city architecture.

pressure, and other types of information can all be collected. For capturing and delivering data streams to a base station or a sink node for processing, the distributed sensor grid is integrated into municipal infrastructure. Data routing and transmission to the service-oriented middleware layer are handled by wired and wireless network devices. Many studies [7], [14]–[16], consider this layer as two separate layers, the sensors layer, and the network layer.

2. Data storage, real-time processing, and analysis are handled by the service-oriented middleware layer. To promote interoperability across otherwise incompatible technologies, service-oriented architecture often uses widely recognized Internet standards such as REST or Web Services. [17]. Data can be preprocessed with very low latency at edge gateways near smart devices, and then aggregated and processed in data centers using cloud computing and machine learning capabilities. [18].

3. The intelligent application layer provides end-users with a user interface that is efficient, interactive, and comprehensive, allowing intelligent services for many city domains to be provided. Tables, charts, and handlers should be used to provide information to the user in a clear and comprehensible manner, allowing him to experiment and create scenarios. The application layer should offer the user all the tools they need to make informed decisions and plan their city.

6.2.2 Social media

The use of sensors in IoT applications can support finding "what" is occurring in a city but is unable to detect "why" and "how" this happens. Cities grow and operate like living organisms that dynamically change as a result of how people conduct, act, and live in them. [19]. These procedures are reflected on social networks. A social network is defined as "an organized set of people that consists of two kinds of elements: human beings and the connections between them" [20]. People around the world use social media to connect and interact with each other, sharing and propagating information, ideas, and opinions on different areas of urban life [21]. Thus, social media can be considered a significant source of city information as people move around the city places. Insights on current city conditions can be revealed using Social Media Analysis (SMA). SMA techniques consist of machine learning algorithms along with statistical and mathematical methodologies harnessing social media data [22].

City administrators need to consider the concerns and perceptions of citizens as they can contribute to the overall development of smart cities. Citizen participation is more than ever becoming dominant for smart city solutions, without central planning and control [23]. However, taking opinions from all people can cause chaos in the decision-making process. Novel influence metrics comprise the identification of different roles in the network, such as influential citizens, opinion leaders, topical experts, authoritative actors, and influence spreaders or disseminators [24]. Posts or replies from topical influential nodes on several topics about city areas can be useful for detecting citizen sentiments and trends.

The connections between entities are changing each time a person follows another person or makes new "friends". They are also changing when someone replies or reacts to a post. The appearance of a new user or a new post can make dozens of new connections. The intuitive way to represent such connections is using a graph [25].

6.2.3 Knowledge graphs

In the 18th century, the Swiss mathematician L. Euler was inspired by the "Königsberg Brückenproblem" to invent graph theory. The problem aimed to find the shortest route that the emperor of Prussia would have to follow to visit all the land areas of the city and return to the starting point without crossing the same bridge a second time. To address the problem, Euler created an abstract map of the area, showing only the components that affect the problem, as shown in Figure 6.4. He represented every land area as a node and every bridge as an edge. Euler proved that the problem did not need to involve the physical geography of the area and that the important criterion was the existence of an even number of bridges in one of the land areas of the city so that the emperor could start his journey from there. As there was no such area, Euler concluded that there was no solution to the problem that met the predetermined restrictions [26]. Since then, graph theory has

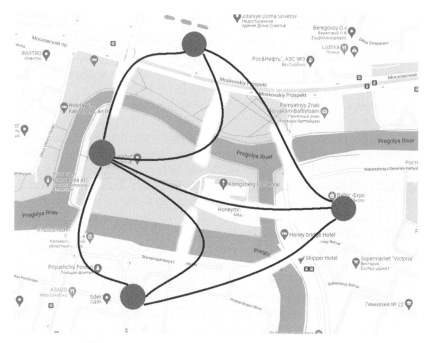

Figure 6.4. A graph as an abstract map of Königsberg.

grown and evolved into an efficient tool in the field of applied mathematics. Today it finds many applications in various scientific areas, such as biology, chemistry, sociology, engineering, computer science, and operations research.

A knowledge graph is a specific type of graph made of "entity-relation-entity" triples and the value pairs of entities and their associated attributes [27]. It connects the fragmented knowledge by representing a semantic network that can effectively support the contextual understanding of a domain. Knowledge graphs were originally proposed by Google in 2012 as a way of improving search procedures. With a knowledge graph, the intricate relationships between the real-world entities can be inferred out of existing facts [28] and complex related information can be represented in a structured way [29]. As Qamar et al. [30] state, having all the data semantically structured not only displays the domain's structure in a comprehensible way but also allows academics to work on the data without exerting additional effort. Knowledge graphs' holistic understanding can assist in extracting insights from existing data, improving predictions, and driving automation and process optimization [26].

The benefits and applications of knowledge graphs for smart cities are not been explored yet [31]. Decision-makers must deal with the complexity and variety of municipal data by utilizing technologies that are aware of the smart city pillars' interdependencies [32]. The use of relational tables or key-value pairs is very common in traditional smart city applications. It is difficult to retrieve qualitative information from these databases, because of the high volume and fragmentation of the stored data [33]. There is a need to make data smarter. This may be achieved by combining data and knowledge at a large scale, which is why knowledge graphs are developed. [26].

Kurteva and Fensel [31] argue that knowledge graphs can provide the level of traceability, transparency, and interpretability that is needed for the implementation of smart city holistic management as they are widely used in knowledge-driven tasks, such as information retrieval, question-answering, intelligent dialogue systems, personalized recommendations, and visualization [27, 34]. City data are expected to be understood more intuitively and better serve the decision-making process with the use of a visual knowledge graph. There is a direct correspondence between the three layers of smart city architecture and the knowledge graphs taxonomy that Aasman [4] proposed. Aasman classifies knowledge graphs into three categories:

- Internal operations knowledge graph: It focuses on integrating an organization's human resources, materials, facilities, and projects. This knowledge graph yields a competitive advantage to the organization by improving its self-knowledge.
- Intermediary products and services knowledge graph: It is dedicated to the enterprise industry, line of business, and operation sector and is meant to improve services.
- External customer knowledge graph: This graph integrates data coming from various organization silos. It focuses on the relationships between these pieces of information, offering the typical 360-degree view of an organization's customers.

An ontology describes the top-level structure of a knowledge graph. It is a taxonomy scheme that identifies the classes in a domain and the relationships between them. An ontology can be constructed from a knowledge graph for system analysis and level of domain knowledge detection [26, 29]. Komninos et al. [23] argue that the impact of smart city applications depends primarily on their ontology, and secondarily on technology and programming features.

6.2.4 Graph databases

In recent years, graph databases have become popular alternatives to the traditional relational model due to their advantages in handling data elements that have intricate relationships and their flexibility, scalability, and high-performance [35, 36]. As graph databases are designed to handle highly connected data effectively, they are suited for cross-domain analysis as the core component of a knowledge management system in several applications, such as social networks, recommendation systems, and fraud detection [37, 38]. Graph databases are also efficient and convenient to handle IoT data as they enable the representation of a vast amount of interconnections between nodes on a network [25].

Graph databases support Creating, Retrieving, Updating, and Deleting (CRUD) facilities [39] and they guarantee reliability as they comply with atomicity, consistency, isolation, and durability (ACID) properties [38]. Beyond just the database system itself, graph databases also comprise graph query languages like Gremlin or Cypher, which support associated query and graph algorithms [40]. Graph algorithms accomplish understanding the relationships and structure within a network by examining the overall topology of the system through its connections [41].

With graph databases, query performance tends to be relatively constant, even as the dataset gets bigger, in contrast to relational databases, where performance declines as the dataset grows [36]. Ding et al. [42] conducted extensive experiments and found that graph databases are shown to be superior to relational databases for projection, multi-table join, and deep recursive operations, but relational databases perform better for aggregations and order by operations. The difference in performance when dealing with highly connected data is achieved by the ability of the graph databases to focus locally on a part of the network and not on the whole dataset.

The advantages of graph databases are not limited to performance issues, as they also show compelling flexibility. With graph databases, data modeling differs from other modeling techniques [33]. Unlike relational databases, there is no need to define traditional tables and fields before adding data. The real-world entities are represented by nodes that can have attributes and we can define semantically relevant relationships between any two nodes to associate them. The schema-free nature of graph databases allows adapting a data model to evolving organization requirements [40] by adding new nodes, new labels, new kinds of relationships, and new subgraphs to an existing structure without disturbing application functionality and existing queries [36].

In this chapter, the Neo4j graph database is the proposed option for the storage, management, and visualization of a smart city knowledge graph. According to [43], Neo4j is considered the most popular graph database. Neo4j not only employs the visualization of a knowledge graph but also effectively stores the information in its graph data structure so that we can directly run queries [33, 44]. Gorawski and Grochla [45] state that Neo4j stores a structure similar to physical infrastructure and blends virtual nodes with real data about physical objects to accommodate a digital representation of the physical city.

6.3   Materials and methods

This study aims to examine how smart city-linked data can generate new knowledge by employing the capabilities of the emerging technology of graph databases. An integrated literature evaluation was conducted to meet the study's objectives. An integrative literature review is a kind of study that generates new information by combining concepts from the literature to create a composite model about a topic.

It creatively explores, evaluates, and synthesizes representative literature on a topic to produce new frameworks and perspectives on the subject [46].

The current study is organized under the approach that is outlined in [47]. First, combinations of related keywords consist of "graph databases", "knowledge graphs", "smart cities", and "social media", included in the queries posed to several academic databases such as Scopus, Google Scholar, and Eric. The search outcomes returned hundreds of articles published in journals, conference proceedings, and books so we had to filter the results through a process of critical reading of their abstracts to ensure the maximum possible relevance to the subject. The collection of the most relevant articles consisted of a set of 73 studies that had been published from 2013 to 2021. A detailed review was then performed to form a comprehensive representation of the cognitive structure of the concepts that make up the field. The next step involved synthesizing the ideas found in the relevant literature by looking for the links that connect these concepts. A proposition of an integrative framework for managing smart city-linked data with graph databases is derived from the synthesis of the literature. The whole process is illustrated in Figure 6.5.

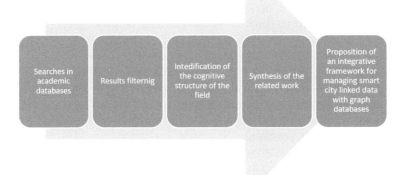

Figure 6.5. Current study process.

6.4   Overview of the literature

There are three main views on how graph databases are utilized in smart city management. The social media application view includes works where graph databases are used for storing and querying data collected from social media. The IoT applications view comprises studies where graph databases support storing and querying IoT applications data. Finally, the integrating view includes works that focus on aggregating several heterogeneous smart city applications in an embedded system.

Tidke et al. [24] state that the identification of influential nodes in a social network can act as a source in making public decisions because their opinions give insights to urban governance bodies. For this purpose, they present novel approaches to identify and rank influential nodes for the smart city topic based on a timestamp. Their approach differs from most state-of-the-art approaches because it can be stretched to test different heterogeneous features on diverse data sets and applications.

In [48] a novel model for autonomous citizen profiling capable to resolve intelligent queries is proposed. The researchers have used sousveillance and surveillance devices to provide the information that is represented as a knowledge graph for inferring and matching relationships among the entities. Tsitseklis et al. [49] suggested a community detection approach to analyze complex networks that use hyperbolic network embedding to obtain communities in very large graphs. The system is based on a graph database for storing information as well as supporting crucial computations.

Wang et al. [50] proposed a heterogeneous graph embedding framework for a location-based social network. As location-based social networks are heterogeneous by nature because they contain various types of nodes, i.e., users, Points Of Interest (POIs), and various relationships between different nodes, they argue that their framework can extract and represent useful information for smart cities. D'Onofrio et al. [51] investigated how a fuzzy reasoning process with multiple components (such as natural language processing and information retrieval) could improve information processing in urban systems. They modeled city data by fuzzy cognitive maps stored in graph databases.

The following papers on IoT applications for city transportation were identified during the review of the relevant literature. Wirawan et al. [39] proposed a database of multimodal transportation design using the graph data model. In [52] a graph-based framework that helps to identify the impact of breakage on very large road networks is presented. The authors analyze vulnerability using global efficiency, a metric that has been widely used to characterize the overall resilience of several city networks such as power grids and transportation networks. Then, to identify the most vulnerable nodes inside the network, they use betweenness centrality, a fundamental metric of centrality to identify topological criticalities.

Vela et al. [53] aim to discover the strengths or weaknesses of the public transport information provided and the services offered by utilizing the Neo4j graph database. In their study, they propose a novel method to design a graph-oriented database in which to store accessible routes generated by mobile application users. Tan et al. [29] created a representation learning model (named TransD) to perform knowledge reasoning using the existing knowledge in knowledge graphs, which can discover the implicit relationship between traffic entities, such as the relationship between the POIs and the road traffic state. The objective of their study was to construct the domain ontology with the four elements of "people–vehicle–road–environment" in traffic as the core, and related to traffic subjects, travel behavior, traffic facilities, traffic tools, and other entities.

Bellini et al. [54] suggested a system for the ingestion of public and private data for smart cities, including road graphs, services available on the roads, traffic sensors, and other relevant features. The system can handle enormous amounts of data from a variety of sources, including both open data from government agencies and private data.

A wide range of studies is also found on the topic of energy management in cities with the help of IoT and graph database technologies. Huang et al. [34] used an AI-enhanced semi-automated labeling system to construct a knowledge graph model for facilitating the grid management and search functions with the help of Neo4j.

Kovacevic et al. [55] elaborate on four different approaches which aim to tighten access control to preserve privacy in smart grids. They also employed the Neo4j graph database for storing and querying the power distribution network model in their study. Graph-based models can be used also for handling big data generated from surveillance applications in wireless multimedia sensor networks as the authors did in [37]. In their study, big sensor data is stored in a well-defined graph database for simulating multimedia wireless sensor networks to run several complex experimental queries.

Gorawski & Grochla [45] proposed a graph database schema for representing the linear infrastructure of the city, e.g., railways, waterworks, canals, gas pipelines, heat pipelines, electric power lines, cable ducting, or roads for identifying the real connections between these linear infrastructure objects and connect them with devices, meters, and sensors.

Le-Phuoc et al. [56] created a knowledge graph to pave the way toward building a "real-time search engine for the Internet of Things", which they call the Graph of Things (GoT), which aims at enabling a deeper understanding of the data generated by connected things of the world around us. Palaiokrassas et al. [38] built a Neo4j graph database for storing data collected from smart city sensors (about air pollution concentrations, temperature, and relative humidity) as well as city open data sources and employed recommendations for citizens. D'Orazio et al. [57] show how developing techniques for querying and evolving graph-modeled datasets based on user-defined constraints can be applied to effectively create knowledge from urban data with automated mechanisms that guarantee data consistency.

To solve the problem of multi-source spatiotemporal data analysis in heterogeneous city networks, Zhao et al. [58] proposed a general framework via knowledge graph embedding for multi-source spatiotemporal data analysis tasks. They then used link prediction and cluster analysis tasks to mine the network structure and semantic knowledge. Ali et al. [59] proposed a particularly customized to a smart city environment Semantic Knowledge-Based Graph (SKBG) model as a solution that overcomes the basic limitations of conventional ontology-based approaches. Their model interlinks heterogeneous data to find meaning, concepts, and patterns in smart city data.

Consoli et al. [60] outline a general methodology for developing consumable semantic data models for smart cities. Their approach produces a clean city-specific ontology by leveraging well-known open standards, using extensive metadata, and pursuing semantic interoperability at the domain level. Psyllidis [3] introduced OSMoSys, a knowledge representation framework for smart city planning and management that enables the semantic integration of heterogeneous urban data from diverse sources. By providing tools that can operate across heterogeneous data silos, the suggested framework promotes mutual engagement among urban planners, city managers, decision-makers, and other city stakeholders.

Maduako and Wachowicz [61] proposed a space-time varying graph built on the whole-graph method, where a network structure grows in time and space in such a way that evolutionary patterns are due to the changes in the connectivity and adjacency relationships among network nodes. Graph database systems take advantage of data relationships and facilitate knowledge extraction concerning urban entities and their connections, particularly when combined with spatiotemporal data to identify events or conditions that occur in the city environment [62].

Schoonenberg et al. [63] showed that hetero-functional graph theory can be employed in arbitrary topologies of interdependent smart city networks. The theory can accommodate as many layers (heterogeneous infrastructure networks) as required in the analysis. There is plenty of related work on integrating independent smart city applications into an embedded model. Yao et al. [64] developed the '3D City Database', an open-source 3D geo-database solution for 3D city model development and deployment.

Fernandez et al. [65] proposed a hybrid model based on augmented reality (AR) and augmented space (AS) principles. The proposed framework uses a graph database and a Geographic Information System (GIS) and allows managing information about city services, environment, and context. Qamar et al. [30] presented Smart City Service Ontology (SCSO) that can be used for all smart city applications as they argue. SCSO provides a centralized model for smart city applications that improves data management by incorporating semantic knowledge. They demonstrated four different smart city applications, including a smart complaint management system, a smart parking system, a smart street light management system, and smart garbage bin control using SCSO, to demonstrate its ability to integrate various applications into a single integrated platform.

Ontologies have been widely used to represent the conceptual framework of smart cities. Přibyl et al. [66] argue that an ontology needs to be prepared by an expert (or rather a group of experts). There is no one, who can be considered an expert in all various domains of smart cities. Štěpánek and Ge [67] designed a smart city ontology by fitting a set of critical papers selected from the literature. Liu et al. [27] designed a method for analyzing smart city data by using the ontology of the smart city index system.

Qamar and Bawany [68] proposed a layered architecture for smart city security by employing an ontological framework for smart city application, communication, and data management. Smirnova and Popovich [69] presented a new approach to the development of an ontology of the city environment which implies the combination of two ontology models: a) ontology of the subject domain and b) scenario ontology. This would allow not only for the organization of information and knowledge about the city environment but also for the improvement of the quality of services offered by the city environment at all management levels by developing simulations of probable scenarios involving city environment objects.

The seven requirements for data management in smart city applications: data heterogeneity, semantic interoperability, scalability, real-time processing, security, spatial data handling, and data aggregation that Amghar et al. [70] consider, create specific demands for the features of the databases to be used. Effendi et al. [71] proved that for spatiotemporal data, graph database technology outperforms relational databases due to the interconnected nature of the data.

Brugrana et al. [72] provided insights into theoretical strong and weak spots for seven graph database systems. In their study, they analyzed four scenarios in the context of smart cities and suggested the best solution for each use case. Almabdy [73] conducted a comparative study between a graph and a relational database and states that the former is the preferred technology for smart city applications as it offers better graph structures and it is flexible and white-board friendly. Desai et al. [74] explored various data, storage, and indexing models used for graph databases for smart city applications by elaborating on their main features, strengths, and weaknesses.

6.5   Synthesis and discussion

The knowledge acquired from the detailed and critical review of the literature enables a better understanding of the field through synthesis. Synthesis reorganizes, combines, and enhances existing concepts with new ideas to create new ways of thinking about a subject [46]. The literature has shown that a city is a complex system that consists of interacting layers as illustrated in Figure 6.6. Each layer will be more closely linked with other layers, forming a closed-loop system [27]. More specifically, a city includes:

- The geospatial layer: City networks do not simply link entities, rather they connect urban places. Geospatial data are by nature large in quantity and heterogeneous in quality [75]. Although the special characteristics of geospatial data, specifically multi-dimensionality and the large size of datasets, make processing them with specialized solutions for geo-related analysis [76], graph databases can be an effective alternative solution for storing multimodal geospatial data [39].
- The infrastructure layer: It may include water and sewage networks, power distribution network, telecommunication networks, transportation networks, etc. Each network has multiple subnetworks, for example, the transportation

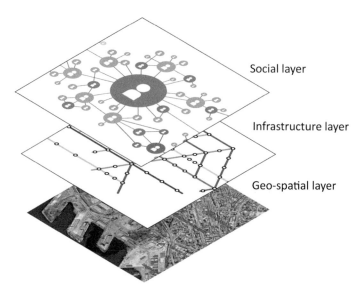

Figure 6.6. The city layers.

network comprises the bus network, the subway network, and the railway network [77]. These networks are not constructed by only one type of technology, but several, and this fact makes their management a particularly big challenge [25].

• The social layer: It may also contain multiple networks, for example, family, friendship, political, business, innovation, academic networks, etc. Mention, however, should be made of social media e.g., Facebook or Twitter, which have become widespread in recent years. Social media have emerged as an important source of information about what is happening in the city [78] as many posts have an associated location. This can lead to a strong correlation between the behavior of social media users and urban spaces [50].

The complexity of city networks and the ambiguities related to social aspects are crucial challenges to the issue of city operation and design [79]. The examination of all the city's components that affect the city and the investigation of the related data flows between these components is vital to create a representative model of the city's structure [69]. The most effective way of representing these networks is graphed where city components are modeled with nodes and the relationships between components are modeled with edges. In addition, each node has to be located in specific geographical coordinates to make sure that all the right things take place at the same location [77]. For example, a streetlight equipped with a Wi-Fi hotspot must be placed near a bus station.

The greatest challenge in integrating smart city applications is interoperability, which arises from data heterogeneity. The hardware and software components that are used in smart city applications are heterogeneous and usually delivered by a variety of suppliers [66]. These applications generate data from measurements about air quality, traffic, transport, energy distribution systems, social events, etc. [62].

Consoli et al. [60] identified a set of data formats that are commonly used for smart city applications. There can be data collected from GIS, data in JSON format from REST web services, XML files, relational databases, Excel files, or graph databases. The integration of such city data sources and the optimized management of city networks needs a multidimensional infrastructure of connectivity [80]. Doing so enables cities to transition from managing data in obsolete independent silos to linked data repositories [4].

These challenges need to be addressed in an integrated approach, ideally by developing (and validating) a framework that is capable of managing complex multi-layer systems comprehensively [81]. Since graphs are a natural way to represent connected systems, graph-oriented databases are emerging as a viable alternative to traditional relational databases [55]. A graph-based data model supports various urban use cases like finding the shortest route (in the geospatial layer), identity and access management (in the infrastructure layer), and real-time recommendation engines (in the social layer) [40]. In addition, the wide range of formats supported by graph databases makes them the ideal tool for integrating city data that are retrieved from different sources [57, 60] into a knowledge graph as shown in Figure 6.7.

A smart city knowledge graph is built on integrated, contextualized, quality-assured, and well-governed data [26]. The characteristics of graph databases, such as schema-free structure, self-explanatory, diversified modeling, domain generality, and implicit storage of relations [74] make them the only pragmatic way of describing the interconnected structure of a smart city knowledge graph. A smart city knowledge graph provides a unified view of city data by combining data stored in multiple information silos. The granular city data coming from different sources provide new possibilities when fused into an integrated graph data model and analyzed together [4].

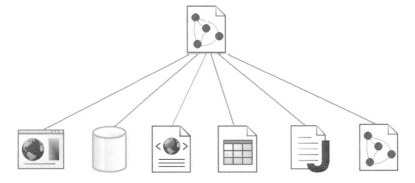

Figure 6.7. Smart city data integration into a knowledge graph.

Angles and Gutierrez [82] define the graph data model as "a model in which the data structures for the schema and/or instances are modeled as a directed, possibly labeled, graph, or generalizations of the graph data structure, where data manipulation is expressed by graph-oriented operations and type constructors, and appropriate integrity constraints can be defined over the graph structure". Küçükkeçeci and Yazıcı [37] consider graph data modeling as a natural way to depict actuality. Robinson et al. [36] underline the fact that graph data models not only represent how things are linked but also provide querying capabilities so to make new knowledge. They argue that graph models and graph queries are just two sides of the same coin.

Needham and Hodler [41] compile a list of graph analytics and algorithms employed for answering questions in graph databases. Some types of algorithms and their related usage are included in Table 6.1. The use cases mentioned in the table are indicative as the possibilities are unlimited.

As we have seen, graph databases can communicate and cooperate effectively with other data technologies, due to their openness and flexibility. However, there is also the issue of interaction with people. The massive volume, high speed, and variety, as well as the unstructured nature of smart city data, cause an intractable problem of making sense of it all.

Managing the entire data ecosystem needs a comprehensive user interface that intuitively provides information. The visual representation of the information has to be dynamic and interactive, and it should be understandable and usable, as it should be delivered to the ordinary citizen and not only to experts. In addition, the user must have the ability to pose queries without any coding knowledge.

A graphic user interface that meets the above requirements provides a better understanding of the entire data ecosystem, from the production of a data point to its consumption, and supports an effective decision-making [26]. Graph databases by nature have an advantage in the data visualization [71]. In contrast to relational databases, which can store data processed and re-shaped into a graph structure, they do not require extra overhead and configuration as this is the native topology of their technology [71]. Graph database visualization mechanisms enable users to intuitively point and click, drag and drop, and query data to obtain information in near real-time and get answers to questions they would not otherwise think to ask [4].

6.6 Conclusion and future work

A city is a complex multilayer system of systems that have evolved as a result of a combination of numerous minor and major human activities. Smart cities, which are built on emerging technologies, such as the IoT, social media, and big data, have revolutionized the way we live and work. They are characterized by high connectivity among the entities that exist in them. Graph databases are an

Table 6.1. Types of graph algorithms and related smart city use cases

Type of algorithm	Usage	Smart city use case
Shortest Path	Finding the optimal path between two nodes depending on the number of hops or any other weighted connection value.	Finding directions between locations.
All Pairs Shortest Path	Understanding alternative routing when the optimal route is blocked or becomes suboptimal.	Optimizing the location of city facilities and the distribution of goods.
Single Source Shortest Path	Evaluating the optimal route from a predetermined start point to all other distinct nodes.	Detecting changes in topology, such as link failures, and suggesting a new routing structure in seconds.
Minimum Spanning Tree	Finding the best route to visit all nodes.	In an outbreak, tracing the history of infection transmission.
Degree Centrality	Finding the "popularity" of individual nodes.	Identifying influential people based on their affiliations, such as social network connections.
Closeness Centrality	Detecting nodes that can spread information efficiently through a subgraph.	Detecting propagation through all shortest paths simultaneously, such as infections spreading through a city.
Betweenness Centrality	Detecting the degree to which a node influences the flow of information or resources in a graph.	Identifying critical transfer nodes in networks like electricity grids.
PageRank	Measuring the number and quality of incoming relationships to a node to determine an estimation of how important that node is.	Predicting traffic flow and human mobility in public spaces.
Triangle Count	Determining the stability of a group.	Investigating the community structure of a social media graph.
Label Propagation	Finding communities in a graph.	Assigning polarity of social media posts as a part of semantic analysis.
Louvain Modularity	Discovering clusters by comparing community density as nodes are assigned to different groups.	Detecting cyberattacks.
Link Prediction	Estimating the likelihood of a relationship forming in the future, or if one should already be there in our graph but isn't owing to missing data.	Making recommendations.

emerging technology that has recently become increasingly popular, proving to be the ideal solution for hosting and managing complex interconnected systems.

To date, implemented smart city applications use different technologies in terms of hardware and software and produce heterogeneous data, stored in independent repositories. Many of these standalone applications have been implemented with graph databases. However, the most important advantage of graph databases is that they can communicate and trade with independent repositories, assuming the role of a metadata hub for the fusion of heterogeneous data, in a holistic ecosystem to act as a city-wide knowledge graph.

Modern graph databases, such as Neo4j, support specialized query languages and graphical algorithms that facilitate getting answers to questions that would not be possible without the interconnection of independent storage silos. Due to the locality of the queries, the performance of the graph algorithms is much higher than that of traditional relational databases. For future research, we plan to implement a mobile application based on graph database technology with an interactive user interface to support informed decisions for citizens. At the same time, the application will allow citizens to import useful information about the situations and events that occurs in the city to enhance and update the smart city knowledge graph.

References

[1] Townsend, A. 2013. Smart Cities: Big Data, Civic Hackers, and the Quest for a New Utopia. New York, NY: W.W. Norton & Company.

[2] Eifrem, E. 2018. The role of graph in our smart city future. *Global*, pp. 7–8.

[3] Psyllidis, A. 2015. Ontology-based data integration from heterogeneous urban systems: A knowledge representation framework for smart cities. *In:* CUPUM 2015 – 14th International Conference on Computers in Urban Planning and Urban Management, July 2015, p. 240.

[4] Aasman, J. 2017. Transmuting information to knowledge with an enterprise knowledge graph. *IT Prof.*, 19(6): 44–51, doi: 10.1109/mitp.2017.4241469.

[5] Goldsmith, S. and S. Crawford. 2014. The Responsive City: Engaging Communities Through Data-Smart Governance. San Francisco, CA, USA: Jossey-Bass.

[6] Charalabidis, Y., C. Alexopoulos, N. Vogiatzis and D. Kolokotronis. 2019. A 360-degree model for prioritizing smart cities initiatives, with the participation of municipality officials, citizens and experts. *In:* M.P.R. Bolivar and L.A. Munoz (Eds.). E-Participation in Smart Cities: Technologies and Models of Governance for Citizen Engagement. pp. 123–153. Cham, Switzerland: Springer International Publishing.

[7] Li, D.R., J.J. Cao and Y. Yao. 2015. Big data in smart cities. *Sci. China Inf. Sci.*, 58(10): 1–12, doi: 10.1007/s11432-015-5396-5.

[8] Ejaz, W. and A. Anpalagan. 2019. Internet of Things for Smart Cities: Technologies, Big Data and Security. Cham, Switzerland: Springer Nature.

[9] D'Aniello, G., M. Gaeta and F. Orciuoli. 2018. An approach based on semantic stream reasoning to support decision processes in smart cities. *Telemat. Informatics*, 35(1): 68–81, doi: 10.1016/j.tele.2017.09.019.

[10] Kousis, A. and C. Tjortjis. 2021. Data Mining Algorithms for Smart Cities: A Bibliometric Analysis. *Algorithms, 2021*, 14(8): 242, https:// doi.org/10.3390/ a14080242

[11] Pieroni, A., N. Scarpato, L. Di Nunzio, F. Fallucchi and M. Raso. 2018. Smarter city: Smart energy grid based on blockchain technology. *Int. J. Adv. Sci. Eng. Inf. Technol.*, 8(1): 298–306, doi: 10.18517/ijaseit.8.1.4954.

[12] Khan, M.S., M. Woo, K. Nam and P.K. Chathoth. 2017. Smart city and smart tourism: A case of Dubai. *Sustain.*, 9(12), doi: 10.3390/su9122279.

[13] Kumar Kar, A., S.Z. Mustafa, M. Gupta, V. Ilavarsan and Y. Dwivedi. 2017. Understanding smart cities: Inputs for research and practice. *In:* A. Kumar Kar, M.P. Gupta, V. Ilavarsan and Y. Dwivedi (Eds.). Advances in Smart Cities: Smarter People, Governance, and Solutions. pp. 1–8. Boca Raton, FL: CRC Press.

[14] Li, D., J. Shan, Z. Shao, X. Zhou and Y. Yao. 2013. Geomatics for smart cities – Concept, key techniques, and applications. *Geo-Spatial Inf. Sci.*, 16(1): 13–24, doi: 10.1080/10095020.2013.772803.

[15] Moreno, M.V., F. Terroso-Sáenz, A. González-Vidal, M. Valdés-Vela, A, Skarmeta, M.A. Zamora and V. Chang. 2017. Applicability of big data techniques to smart cities deployments, *IEEE Trans. Ind. Informatics*, 13(2): 800–809, doi: 10.1109/ TII.2016.2605581.

[16] Massana, J., C. Pous, L. Burgas, J. Melendez and J. Colomer. 2017. Identifying services for short-term load forecasting using data driven models in a Smart City platform. *Sustain. Cities Soc.*, 28: 108–117, doi: 10.1016/j.scs.2016.09.001.

[17] Wlodarrczak, P. 2017. Smart Cities – Enabling technologies for future living. *In:* A. Karakitsiou, A. Migdalas, S. Rassia and P. Pardalos (Eds.). City Networks: Collaboration and Planning for Health and Sustainability. pp. 1–16. Cham, Switzerland: Springer International Publishing.

[18] Mystakidis, A. and C. Tjortjis. 2020. Big Data Mining for Smart Cities: Predicting Traffic Congestion using Classification. doi: 10.1109/IISA50023.2020.9284399.

[19] Giatsoglou, M., D. Chatzakou, V. Gkatziaki, A. Vakali and L. Anthopoulos. 2016. CityPulse: A platform prototype for smart city social data mining. *J. Knowl. Econ.*, 7(2): 344–372, doi: 10.1007/s13132-016-0370-z.

[20] Christakis, N. and J. Fowler. 2011. Connected: The Amazing Power of Social Networks and How They Shape Our Lives. London, UK: Harper Press.

[21] Koukaras, P., C. Tjortjis and D. Rousidis. 2019. Social Media Types: Introducing a Data Driven Taxonomy, vol. 102(1). Springer Vienna.

[22] Rousidis, D., P. Koukaras and C. Tjortjis. 2019. Social Media Prediction: A Literature Review, vol. 79(9–10). Multimedia Tools and Applications.

[23] Komninos, N., C. Bratsas, C. Kakderi and P. Tsarchopoulos. 2015. Smart City Ontologies: Improving the effectiveness of smart city applications. *J. Smart Cities*, 1(1): 31–46, doi: 10.18063/jsc.2015.01.001.

[24] Tidke, B.A., R. Mehta, D. Rana, D. Mittal and P. Suthar. 2021. A social network based approach to identify and rank influential nodes for smart city. *Kybernetes*, 50(2): 568–587, doi: 10.1108/K-09-2019-0637.

[25] CITO Research. 2020. Six Essential Skills for Mastering the Internet of Connected Things. Retrieved April 26, 2022 from http://www.citoresearch.com

[26] Barrasa, J., A.E. Hodler and J. Webber. 2021. Knowledge Graphs: Data in Context for Responsive Businesses. Sebastopol, CA: O'Reilly Media.

[27] Liu, J., H. Ning, Y. Bai and T. Yan. 2020. The study on the index system of the smart

city based on knowledge map. *J. Phys. Conf. Ser.*, 1656(1), doi: 10.1088/1742-6596/1656/1/012015.

[28] Fensel, A. 2020. Keynote: Building smart cities with knowledge graphs. 2019 Int. Conf. Comput. Control. Informatics its Appl., 19: 1–1, doi: 10.1109/ic3ina48034.2019.8949613.

[29] Tan, J., Q. Qiu, W. Guo and T. Li. 2021. Research on the construction of a knowledge graph and knowledge reasoning model in the field of urban traffic. *Sustain.*, 13(6), doi: 10.3390/su13063191.

[30] Qamar, T., N.Z. Bawany, S. Javed and S. Amber. 2019. Smart City Services Ontology (SCSO): Semantic Modeling of Smart City Applications. *In:* Proc. 2019 7th Int. Conf. Digit. Inf. Process. Commun. ICDIPC, pp. 52–56, doi: 10.1109/ICDIPC.2019.8723785.

[31] Kurteva, A. and A. Fensel. 2021. Enabling interpretability in smart cities with knowledge graphs: Towards a better modelling of consent. *IEEE Smart Cities*, https://smartcities.ieee.org/newsletter/june-2021/enabling-interpretability-in-smart-cities-with-knowledge-graphs-towards-a-better-modelling-of-consent (accessed Aug. 17, 2021).

[32] De Nicola, A. and M.L. Villani. 2021. Smart city ontologies and their applications: A systematic literature review. *Sustain.*, 13(10), doi: 10.3390/su13105578.

[33] Huang, H., Y. Chen, B. Lou, Z. Hongzhou, J. Wu and K. Yan. 2019. Constructing knowledge graph from big data of smart grids. Proc. 10th Int. Conf. Inf. Technol. Med. Educ. ITME 2019, pp. 637–641, doi: 10.1109/ITME.2019.00147.

[34] Huang, H., Z. Hong, H. Zhou, J. Wu and N. Jin. 2020. Knowledge graph construction and application of power grid equipment. Math. Probl. Eng., vol. January 2020, no. 2018, doi: 10.1155/2020/8269082.

[35] Zhang, Z.J. 2017. Graph databases for knowledge management. *IT Prof.*, 19(6): 26–32, doi: 10.1109/MITP.2017.4241463.

[36] Robinson, I., J. Webber and E. Eifrem. 2015. Graph Databases: New Opportunities for Connected Data, 2nd ed. Sebastopol, CA, USA: O'Reilly.

[37] Küçükkeçeci, C. and A. Yazıcı. 2018. Big Data Model Simulation on a graph database for surveillance in wireless multimedia sensor networks. *Big Data Res.*, 11: 33–43, doi: 10.1016/j.bdr.2017.09.003.

[38] Palaiokrassas, G., V. Charlaftis, A. Litke and T. Varvarigou. 2017. Recommendation service for big data applications in smart cities. *In:* International Conference on High Performance Computing and Simulation, HPCS 2017, pp. 217–223, doi: 10.1109/HPCS.2017.41.

[39] Wirawan, P.W., D. Er Riyanto, D.M.K. Nugraheni and Y. Yasmin. 2019. Graph database schema for multimodal transportation in Semarang. *J. Inf. Syst. Eng. Bus. Intell.*, 5(2): 163–170, doi: 10.20473/jisebi.5.2.163-170.

[40] Webber, J. 2021. The Top 10 Use Cases of Graph Database Technology. [Online]. Available: https://neo4j.com/resources/top-use-cases-graph-databases-thanks/?aliId=eyJpIjoiWlA2MFdod08yeW14bmt0XC8iLCJ0IjoiTEg3aFVUdlwvWHQ3dDgwV1djQUdYUVE9PSJ9.

[41] Needham, M. and A.E. Hodler. 2019. Graph Algorithms: Practical Examples in Apache Spark & Neo4j. Sebastopol, CA, USA: O'Reilly.

[42] Ding, P., Y. Cheng, W. Lu, H. Huang and X. Du. 2019. Which Category is Better: Benchmarking the RDBMSs and GDBMSs. *In:* Web and Big Data: 3rd International Joint Conference, APWeb-WAIM 2019 Proceedings, Part II, vol. 2, pp. 191–206, [Online]. Available: http://link.springer.com/10.1007/978-3-030-26075-0.

[43] DB-Engines Ranking. 2021. DB-Engines Ranking – Popularity ranking of database management systems. https://db-engines.com/en/ranking (accessed Nov. 17, 2021).

[44] Zhao, Y., Y. Yu, Y. Li, G. Han and X. Du. 2019. Machine learning based privacy-preserving fair data trading in big data market. *Inf. Sci. (Ny).*, 478: 449–460, doi: 10.1016/j.ins.2018.11.028.

[45] Gorawski, M. and K. Grochla. 2020. Performance tests of smart city IoT data repositories for universal linear infrastructure data and graph databases. *SN Comput. Sci.*, 1(1): 1–7, doi: 10.1007/s42979-019-0031-y.

[46] Torraco, R.J. 2016. Writing integrative literature reviews: Using the past and present to explore the future. *Hum. Resour. Dev. Rev.*, 15(4): 404–428, doi: 10.1177/1534484316671606.

[47] Torraco. R.J. 2005. Writing integrative literature reviews: Guidelines and examples. *Hum. Resour. Dev. Rev.*, 4(3): 356–367, doi: 10.1177/1534484305278283.

[48] Munir, S., S.I. Jami and S. Wasi. 2020. Towards the modelling of veillance based citizen profiling using knowledge graphs. *Open Comput. Sci.*, 11(1): 294–304, doi: 10.1515/comp-2020-0209.

[49] Tsitseklis, K., M. Krommyda, V. Karyotis, V. Kantere and S. Papavassiliou. 2020. Scalable community detection for complex data graphs via hyperbolic network embedding and graph databases. *IEEE Trans. Netw. Sci. Eng.*, 8(2): 1269–1282, doi: 10.1109/tnse.2020.3022248.

[50] Wang, Y., H. Sun, Y. Zhao, W. Zhou and S. Zhu. 2020. A heterogeneous graph embedding framework for location-based social network analysis in smart cities. *IEEE Trans. Ind. Informatics*, 16(4): 2747–2755, doi: 10.1109/TII.2019.2953973.

[51] D'Onofrio, S., S.M. Müller, E.I. Papageorgiou and E. Portmann. 2018. Fuzzy reasoning in cognitive cities: An exploratory work on fuzzy analogical reasoning using fuzzy cognitive maps. IEEE International Conference on Fuzzy Systems, 2018, vol. July 2018, doi: 10.1109/FUZZ-IEEE.2018.8491474.

[52] Furno, A., N.E. El Faouzi, R. Sharma, V. Cammarota and E. Zimeo. 2018. A graph-based framework for real-time vulnerability assessment of road networks. Proceedings 2018 IEEE International Conference on Smart Computing, SMARTCOMP 2018, pp. 234–241, doi: 10.1109/SMARTCOMP.2018.00096.

[53] Vela, B., J.M. Cavero, P. Cáceres, A. Sierra and C.E. Cuesta. 2018. Using a NoSQL graph oriented database to store accessible transport routes. Workshop Proceedings of the EDBT/ICDT 2018 Joint Conference, pp. 62–66.

[54] Bellini, P., M. Benigni, R. Billero, P. Nesi and N. Rauch. 2014. Km4City ontology building vs data harvesting and cleaning for smart-city services. *J. Vis. Lang. Comput.*, 25(6): 827–839, doi: 10.1016/j.jvlc.2014.10.023.

[55] Kovacevic, I., A. Erdeljan, M. Zaric, N. Dalcekovic and I. Lendak. 2018. Modelling access control for CIM based graph model in Smart Grids. 2018 Int. Conf. Signal Process. Inf. Secur. ICSPIS 2018, pp. 1–4, 2019, doi: 10.1109/CSPIS.2018.8642760.

[56] Le-Phuoc, D., H. Nguyen Mau Quoc, H. Ngo Quoc, T. Tran Nhat and M. Hauswirth. 2016. The Graph of Things: A step towards the Live Knowledge Graph of connected things. *J. Web Semant.*, 37–38: 25–35, doi: 10.1016/j.websem.2016.02.003.

[57] D'Orazio, L., M. Halfeld-Ferrari, C.S. Hara, N.P. Kozievitch and M.A. Musicante. 2017. Graph constraints in urban computing: Dealing with conditions in processing urban data. 2017 IEEE International Conference on Internet of Things, IEEE Green Computing and Communications, IEEE Cyber, Physical and Social Computing, IEEE Smart Data, iThings-GreenCom-CPSCom-SmartData 2017, pp. 1118–1124, doi: 10.1109/iThings-GreenCom-CPSCom-SmartData.2017.171.

[58] Zhao, L., H. deng, L. Qiu, S. Li, Z. Hou, H. Sun and Y. Chen. 2020. Urban multi-source spatio-temporal data analysis aware knowledge graph embedding. *Symmetry (Basel).*, 12(2): 1–18, doi: 10.3390/sym12020199.

[59] Ali, S., G. Wang, K. Fatima and P. Liu. 2019. Semantic knowledge based graph model in smart cities. Communications in Computer and Information Science: Smart City and Informatization, 7th International Conference, iSCI 2019, vol. 1122, pp. 268–278, doi: 10.1007/978-981-15-1301-5_22.

[60] Consoli, S., M. Mongiovic, A.G. Nuzzolese, S. Peroni, V. Presutti, D.R. Recupero and D. Spampinato. 2015. A smart city data model based on semantics best practice and principles. WWW 2015 Companion – Proceedings of the 24th International Conference on World Wide Web, pp. 1395–1400, doi: 10.1145/2740908.2742133.

[61] Maduako, I. and M. Wachowicz. 2019. A space-time varying graph for modelling places and events in a network. *Int. J. Geogr. Inf. Sci.*, 33(10): 1915–1935, doi: 10.1080/13658816.2019.1603386.

[62] Amaxilatis, D., G. Mylonas, E. Theodoridis, L. Diez and K. Deligiannidou. 2020. LearningCity: Knowledge generation for smart cities. *In:* F. Al-Turjman (Ed.). Smart Cities Performability, Cognition & Security. pp. 17–41. Cham, Switzerland: Springer.

[63] Schoonenberg, W.C.H., I.S. Khayal and A.M. Farid. 2018. A Hetero-functional Graph Theory for Modeling Interdependent Smart City Infrastructure. Cham, Switzerland: Springer.

[64] Yao, Z., C. Nagel, F. Kunde, G. Hudra, P. Willkomm, A. Donaubauer, T. Adolphi and T.H. Kolbe. 2018. 3DCityDB – A 3D geodatabase solution for the management, analysis, and visualization of semantic 3D city models based on CityGML. *Open Geospatial Data, Softw. Stand.*, 3(5): 1–26, doi: https://doi.org/10.1186/s40965-018-0046-7.

[65] Fernandez, F., A. Sanchez, J.F. Velez and B. Moreno. 2020. The augmented space of a smart city. International Conference on Systems, Signals, and Image Processing, vol. July 2020, pp. 465–470, doi: 10.1109/IWSSIP48289.2020.9145247.

[66] Přibyl, P., O. Přibyl, M. Svítek and A. Janota. 2020. Smart city design based on an ontological knowledge system. *Commun. Comput. Inf. Sci.*, vol. 1289 CCIS, no. October 2020, pp. 152–164, doi: 10.1007/978-3-030-59270-7_12.

[67] Štěpánek, P. and M. Ge. 2018. Validation and extension of the Smart City ontology. ICEIS 2018 – Proc. 20th Int. Conf. Enterp. Inf. Syst., vol. 2, no. Iceis 2018, pp. 406–413, 2018, doi: 10.5220/0006818304060413.

[68] Qamar, T. and N. Bawany. 2020. A cyber security ontology for Smart City. *Int. J. Inf. Technol. Secur.*, 12(3): 63–74.

[69] Smirnova, O. and T. Popovich. 2019. Ontology-based model of a Smart City. Real Corp 2019: Is this the Real World? Perfect Smart Cities vs. Real Emotional Cities, 2019, pp. 533–540.

[70] Amghar, S., S. Cherdal and S. Mouline. 2018. Which NoSQL database for IoT applications? 2018 Int. Conf. Sel. Top. Mob. Wirel. Networking, MoWNeT 2018, pp. 131–137, doi: 10.1109/MoWNet.2018.8428922.

[71] Effendi, S.B., B. van der Merwe and W.T. Balke. 2020. Suitability of graph database technology for the analysis of spatio-temporal data. *Futur. Internet*, 12(5): 1–31, 2020, doi: 10.3390/FI12050078.

[72] Brugnara, M., M. Lissandrini and Y. Velegrakis. 2016. Graph Databases for Smart Cities. IEEE Smart Cities Initiative. Retrieved September 19, 2021 from https://event.unitn.it/smartcities-trento/Trento_WP_Brugnara2.pdf

[73] Almabdy, S. 2018. Comparative analysis of relational and graph databases for social networks. 1st Int. Conf. Comput. Appl. Inf. Secur. ICCAIS 2018, doi: 10.1109/CAIS.2018.8441982.

[74] Desai, M., R.G. Mehta and D.P. Rana. 2018. Issues and challenges in big graph modelling for Smart City: An extensive survey. *Int. J. Comput. Intell. IoT*, 1(1): 44–50.

[75] Breunig, M., P.E. Bradley, M. Jahn, P. Kuper, N. Mazroob, N. Rösch, M. Al-Doori, E. Stefanakis and M. Jadidi. 2020. Geospatial data management research: Progress and future directions. *ISPRS Int. J. Geo-Information*, 9(2), doi: 10.3390/ijgi9020095.

[76] Guo, D. and E. Onstein. 2020. State-of-the-art geospatial information processing in NoSQL databases. *ISPRS Int. J. Geo-Information*, 9(5), doi: 10.3390/ijgi9050331.

[77] Ameer , F., M.K. Hanif, R. Talib, M.U. Sarwar, Z. Khan, K. Zulfiqar and A. Riasat. 2019. Techniques, tools and applications of graph analytic. *Int. J. Adv. Comput. Sci. Appl.*, 10(4): 354–363, doi: 10.14569/ijacsa.2019.0100443.

[78] Alomari, E. and R. Mehmood. 2018. Analysis of tweets in Arabic language for detection of road traffic conditions. *Lect. Notes Inst. Comput. Sci. Soc. Telecommun. Eng. LNICST*, 224: 98–110, doi: 10.1007/978-3-319-94180-6_12.

[79] Raghothama, J., E. Moustaid, V. Magal Shreenath and S. Meijer. 2017. Bridging borders: Integrating data analytics, modeling, simulation, and gaming for interdisciplinary assessment of health aspects in city networks. *In:* A. Karakitsioy et al. (Eds.). City Networks. pp. 137–155. Cham, Switzerland: Springer International Publishing.

[80] Fontana, F. 2017. City networking in urban strategic planning. *In:* Karakitsiou et al. (Eds.). City Networks. pp. 17–38. Cham, Switzerland: Springer.

[81] Kivelä, M., A. Arenas, M. Barthelemy, J.P. Gleeson, Y. Moreno and M.A. Porter. 2014. Multilayer networks. *J. Complex Networks*, 2(3): 203–271, doi: 10.1093/comnet/cnu016.

[82] Angles, R. and C. Gutierrez. 2008. Survey of graph database models. *ACM Comput. Surv.*, 40(1): 1–39, doi: 10.1145/1322432.1322433.

Graph Databases in Smart City Applications – Using Neo4j and Machine Learning for Energy Load Forecasting

Aristeidis Mystakidis [0000–0002–5260–2906]

The Data Mining and Analytics Research Group, School of Science and Technology, International Hellenic University, GR-570 01 Thermi, Greece
e-mail: a.mystakidis@ihu.edu.gr

The smart city (SC) approach aims to enhance populations' lives through developments in knowledge and connectivity systems such as traffic congestion management, Energy Management Systems (EMS), Internet of Things or social media. An SC is a very sophisticated connected system that generates a big quantity of data and demands a large number of connections. Graph databases (GDB) provide new possibilities for organizing and managing such complicated networks. Machine learning (ML) is positively influencing headline-grabbing apps like self-driving vehicles and virtual assistants, but it is also improving performance and lowering costs for everyday tasks like web chat and customer service, recommendation systems, fraud detection, and energy forecasting. Most of these ML data challenges may be addressed by utilizing GDB. Because graphs are founded on the concept of linking and traversing connections, they are an obvious option for data integration. Graphs can also be used to supplement raw data. Each column in typical tabular data represents one "feature" that the ML system may exploit. Each form of link is an extra feature in a graph. Furthermore, simple graph structures such as causal chains, loops, and forks can be regarded as features in and of themselves. Also, Power system engineers have examined several parallel computing technologies based on relational database structures to increase the processing efficiency of EMS applications but have yet to produce sufficiently rapid results while graph processing does this. Considering Neo4j as the leading NoSQL GDB, the goal of this work is to create and test a method for energy load forecasting (ELF) combining ML and GDB. More specifically, this research integrates multiple approaches for executing ELF tests on historical building data. The experiment produces data resolution for 15 minutes as one step ahead, while disclosing accuracy issues.

7.1   Introduction

For a long time, municipal administration depended on a variety of information sources, including financial analyses, surveys, and ad-hock research, to track the growth of the city and suggest potential growth opportunities. Cities are now outfitted with a plethora of various observation equipment that offers actual input on everything that is going on, due to inexpensive IoT devices, enhanced network connectivity, and breakthroughs in data collection and data analytics. These new streams of data enable authorities to respond strategically and quickly to enhance existing procedures or prevent undesirable situations. The volume, variety, and velocity of data gathered by Smart Cities (SC) are so great that it is referred to as Big Data [1].

There are various SC applications, like healthcare systems [2], traffic congestion management [3], Energy Management Systems (EMS) [4], energy load or generation forecasting [45, 46], waste [5] and water management [18], pollution and air quality monitoring [19] or social media [6]. One of the most important and common aspects of an SC infrastructure that is related to Energy Management Systems with a large variety of publications is Energy Load Forecasting (ELF). ELF is important in power system architecture and control. It supports power providers in estimating energy consumption and planning for anticipated energy requirements. Moreover, it assists transmission and distribution system operators in regulating and balancing upcoming power generation to satisfy energy needs effectively. As a result, even though it could be a challenging task given the complexity of present energy networks, ELF has garnered a great deal of interest in current decades. There is a variety of research related to ELF and EMS with Graph databases (GDB) [7, 8].

One of the most common and known GDB is the Neo4j[1], a GDB that was developed from the ground up to employ both data and connections. Neo4j connects information when it is saved, enabling nearly unprecedented queries at previously unseen rates. Unlike traditional databases, which arrange data in rows, columns, and tables, Neo4j has a flexible structure defined by documented links between data items.

This research analyzes ELF using Machine Learning (ML) time series regression, along with GDB. This entails using conventional evaluation measures, i.e., performance metrics based mostly on statistics, to evaluate the prediction performance of each model. R-squared (R^2), Mean Absolute Error (MAE), Mean Squared Error (MSE), Root Mean Squared Error (RMSE), and Coefficient of

---

[1]   https://neo4j.com/

Variation of Root Mean Squared Error (CVRMSE) are the metrics used. For the initial storage, descriptive analysis, and preprocessing, a Neo4j GDB was employed with the prediction algorithms used in this research can be categorized as ML approaches such as Light Gradient Boosting Machine (LGBM), Random Forest (RF), Artificial Neural Networks (ANN), Extreme Gradient Boosting (XGB) and others. A sliding window method is also used for five distinct prediction models.

In Neo4j, each data record, or node, has direct ties to all of the nodes to which it is related. Because it is built on this simple yet successful optimization, Neo4j performs queries in densely linked data faster and with more depth than other databases.

The remainder of this work has the following structure: Section 2 covers the necessity for reliable ELF, as well as the state-of-the-art and commonly used methods that provide the necessary context. The used principles for time-series ELF are described in Section 3, along with a recommended method that has been created. Section 4 provides the results of the experiments using pilot data. Finally, Section 5 brings the chapter to closure by giving final opinions, conclusions, and future work.

7.2   Background

7.2.1   Graph databases

GDB extend the capabilities of the network and relational databases by providing graph-oriented operations, implementing specific improvements, and providing indexes. Several similar frameworks have lately been built. Despite having comparable qualities, they are produced using diverse technologies and approaches.

The fundamental unit of a GDB is given as G (V, E), where V symbolizes entities and E symbolizes relationships [9]. Singular entity graphs are composed of the same class of nodes, while multiple entity graphs are composed of various types of nodes [9, 10, 11]. Graphs are widely utilized in a variety of real systems due to their adaptable architecture, self-explanatory, diverse modeling, domain universality, schema-free configuration, fast associative functions, and implicit storing of relations [12, 13, 14].

A graph is simply a group of vertices and edges or a series of nodes and the relationships that link them. Graphs describe entities as nodes and the connections between those entities as links [22]. Features, that are crucial pairings, are supported by both nodes or relationships. The information is preserved as a graph in a GDB, and the records are referred to as nodes. Nodes are linked together via relationships. A label is a term that groups nodes together. Neo4j is among the most widespread GDB. The problem would be how can query a graph database, and Neo4j, like any GDB, leverages a sophisticated scientific idea from graph theory as a robust and effective engine for querying data. This is known as graph traversal. Neo4j is an ACID-compliant database that employs Cypher (CQL) as its query language [15, 23]. Neo4j is developed in Java and features about 34 billion nodes and 34 billion interconnections. The following architectural pieces comprise the Neo4j GDB: Properties, Nodes, Relationships, Data Browser, and Labels[2].

7.2.2   Smart cities

The idea of an SC suggests an artificial framework comprised of active monitoring, embedded devices, and decision-making algorithms that analyze real-time data [25]. The goal of an SC is to accomplish effective administration in all sections of the city whilst serving the demands and the requirements of its residents. In addition, it needs to be consistent with the ideals of environmental sustainability, with scientific advancement and collaboration among social and economic stakeholders serving as the primary forces of progress [26].

GDB, ML and big data mining for smart city applications aid in addressing industrial, transportation, and energy management issues in practical ways by incorporating platforms and technologies that effectively move data between apps and stakeholders [27]. Various examples of smart cities use big data mining techniques, such as smart education [28, 29], smart grid [4, 30], and power consumption forecast [4]. As the population rises, so do energy or traffic problems, increasing emissions, and environmental and economic concerns.

7.2.3   Energy load forecasting

Numerous factors influence the relevance of effective electrical load prediction models, the most obvious and significant of which concerns environmental issues. Carbon footprint is among the major factors of environmental changes, and rising rates of CO_2 emission need an immediate remedy. According to the [31], emissions from power generation represent 25% of total global emissions.

Large-scale power storage is costly owing to power transformation losses and the expense of fuel cells. Due to practical limits in storing power, electricity generation, transmission, and distribution should occur concurrently with its need [32]. As a result, power production and demand must preserve a flexible equilibrium. Increased production of electricity could cause energy loss, while increased production could cause a power failure. Precise ELF adds to the supply-demand reliability of the grid [33] and may be used as a benchmark for power system operation and operation often via Demand Response (DR) programs.

Given the time range, ELF may be divided into four major groups [34]. These are very short-term (VSTLF), short-term (STLF), medium-term (MTLF), and long-term load (LTLF).

---

[2] https://neo4j.com/blog/neo4j-3-0-massive-scale-developer-productivity/. Accessed: 2018.05.14.

VSTLF: Very short-term load forecasting is focused on load demand until the upcoming hour. It is primarily employed for real energy framework safety analytics and energy instrument management [34].

STLF: Short-term load forecasting includes from one-hour to one-day or one-week load predictions. STLF is the most useful of the four techniques of load demand forecasting, which can help with systems' energy dispatching and control [35].

MTLF: The purpose of medium-term load forecasting is to predict energy consumption in the coming weeks or months. It is used in the maintenance and operation of power systems [36].

LTLF: Finally, long-term power load forecasting (LTLF) is a method for designing long-term power systems. It is primarily used to anticipate energy consumption for the next year or years [37].

ELF is frequently employed using three unique types of sources. Periodic input factors including load changes induced by air heating and cooling (i.e. quarter of a year, period, weekend, etc.) and prior demand data are instances of these (hourly loads for the previous hour, the previous day, and the same day of the previous week). The anticipated load for every hour of each day, the weekly or monthly or daily power demand, and the daily peak power demand all could be generated by ELF [38].

Furthermore, the author in [39] proposes a network-based deep belief ensemble technique for cooling demand forecasting regarding an air conditioning unit. More precisely, a layered Restricted Boltzmann Machine with a Logistic Regression algorithm is employed as the outcome, with a coefficient of determination (R^2 or R–squared), Mean Absolute Error (MAE), Root Mean Square Error (RMSE), Mean Absolute Percentage Error (MAPE), and Coefficient of Variation Root Mean Squared Error (CVRMSE) evaluators measured.

ANNs [40], tree-based models [41] and Support Vector Machines (SVM) [42] are presently among the most often utilized electricity forecasting methods. SVMs are touted as the most accurate solution for estimating power use according to [43], with levels of complexity and accuracy equivalent to deep neural networks. When compared to ANNs, SVMs have several disadvantages, such as sluggish running speed. This is an issue, particularly with large-scale efforts. Unfortunately, this is a typical issue with moderately realistic models that need a large amount of computer memory and processing or computation time. Moreover, tree-based algorithms could also be among the most efficient models [41], due to their behavior around zero-inflated data sets [44]. Finally, statistical regression methods are a common alternative for forecasting electricity demand. These models are beneficial for determining the significance of prospective model inputs, but they struggle with short-term forecasts because of their high level of inaccuracy [21].

7.3   Methodology

This section details the methods for executing ELF, including the pilot dataset description, preprocessing steps, analysis using Neo4j, and the forecasting technique used, which includes a sliding window design, a common technique for time series regression [2, 20].

7.4   Data collection

The electricity demand from Centre for Research and Technology – Hellas/ Information Technologies Institute (CERTH/ITI) Smart Home[3] was the initial anticipated goal of this study [16]. The data cover a two-year time, commencing on 3/10/2018 and ending on 30/09/2020. The collection contains observations and information from and regarding the CERTH Smart House nano grid system in Thessaloniki, Greece. This facility is a real laboratory for the Centre for Research and Technology-Hellas[4] that was created, installed, and administered by the Information Technology Institute[5].

Furthermore, weather information was also utilized. The information was obtained during a similar time frame as the electricity demand and reflects relative humidity, temperature, wind speed, cloud cover, and timestamps per 15 minutes gap (among other values that were not used). The dataset was obtained from visual crossing and CERTH/ITI[6].

Finally, seasonality parameters produced by the timestamp were utilized, like a weekday, month, day of a year, hour of the day, and quarter of an hour.

7.4.1   Preprocessing and descriptive analysis with Neo4j

The first step in pre-processing was to combine the three data sets using Neo4j. The data were handled with the Neo4j database, with the visualization being conducted employing the GraphXR plugin.

Figure 7.1 presents the aggregated per day Energy (or electricity) demand parameter along with three aggregated per day of several weather parameters (relative humidity, temperature and wind speed) combining the year as the centered parameter. There are three big cycles with each blue center of a cycle representing a specific year (2018, 2019, 2020) of the dataset. Light green bullets represent the wind speed, dark green the temperature, beige the relative humidity and orange the energy load. Bigger bullets illustrate higher values while smaller lower values. Since the only full year of the dataset is 2019 (3/10/2018 to 30/09/2020), the center of the cycle that represents 2019 has 365 (aggregated days) *4 (parameters) + 1 (between the years) connections. Some bullets that are connected with two or

---
[3] https://smarthome.iti.gr/news.html?id=27
[4] www.certh.gr
[5] www.iti.gr
[6] https://www.visualcrossing.com/

three years show that there are daily aggregated measures that have similar values among two or three years.

Figure 7.2 visualizes the temperature with the year as a center showing the green bullets as daily aggregated values. Besides the central connection, there is only one extra aggregated connection, showing that only one daily aggregated temperature value is equal between 2 years.

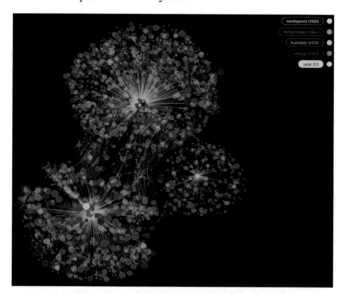

Figure 7.1. Weather and Energy parameters centered with year.

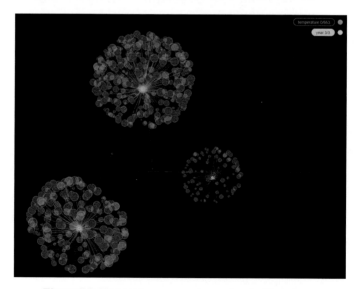

Figure 7.2. Temperature parameter centered with year.

Furthermore, Figure 7.3 illustrates the wind speed parameter centered with the year parameter. As it can be viewed, there are no other connections between the years, besides the central one. This means that there is no daily aggregated wind speed (light green bullets) values that are equal between 2 years. Similarly, Figure 7.4 presents the relative humidity parameter with each year as a center, illustrating the beige bullets as the daily aggregated values. In this case, there are several other connections between the years, besides the central one. This means that several daily aggregated relative humidity values are equal between 2 years.

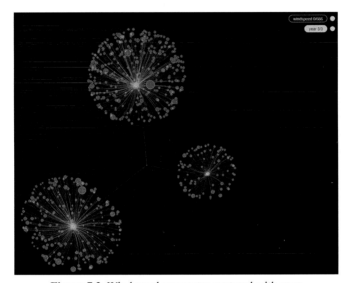

Figure 7.3. Windspeed parameter centered with year.

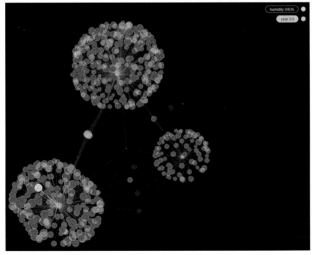

Figure 7.4. Relative humidity parameter centered by year.

Finally, Figure 7.5 shows the graph between energy load and year. Orange bullets show daily aggregated values, while the blue cycle centers the year. As in relative humidity, there is also numerous energy load daily aggregated values that are equal between two years.

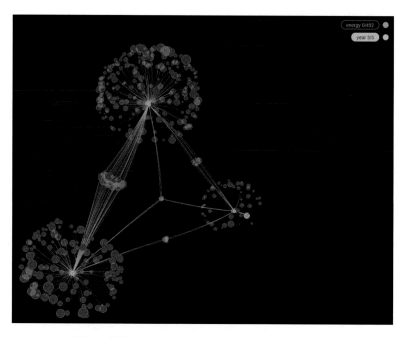

Figure 7.5. Energy load parameter centered by year.

Besides the Neo4j data set combination implementation and descriptive analysis, the final data set comprises a combination of the power measurements, the aforementioned seasonality characteristics, and the platform's meteorological data collection including historical and forecast information. In the case of the latter, a forward linear interpolation strategy was employed to fill the gaps for parameters with fewer than 5% total missing values and no more than eight missing timestamps in a row due to a large number of missing values.

This approach resulted in four complete weather condition parameters, namely Temperature, Wind Speed, Cloud Cover, and Relative Humidity, with the remainder that still had incomplete values excluded from the study. The seasonality parameters used were month, hour, weekday, day of the year, and a quarter of the hour. The combined data set contained nine independent characteristics as well as the power load as the target variable. Furthermore, the histogram from Figure 7.6 illustrates some information regarding the target parameter Energy Load (kWh).

Finally, MinMaxScaler was employed for the network-based models, while for the tree-based no scaling was done.

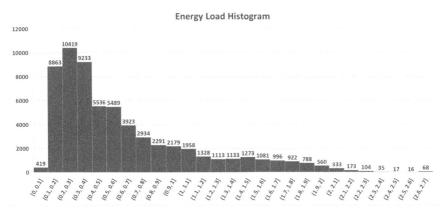

Figure 7.6. Energy load histogram.

7.4.2 ELF with ML

The Sliding Window approach was used to accomplish the time-series regression, comprising a 96-step (previous 24 hours) advance shifting of data, and the result comprises the final data set.

All algorithms used a two-dimension sliding window (time steps with variables and rows) (tree-based and ANN). It entails shifting each characteristic, including the target variable, one to 96 steps ahead of time (one for each quarter of hour) and adding a new column for each step, including also weather forecast parameters for the next timestep of the target parameter. The final data set is the product of this operation.

The data set was therefore subdivided into training and testing parts utilizing a train test split (typical 80%-20% [17], almost 19 months and 6 days for training and five months and 24 days for testing), with the training component being used as an input to all of the ML and DL models. The forecast produced by each algorithm's implementation was compared to the testing portion of the data set and evaluated using some of the most common metrics as mentioned in previous sections. These metrics were MAE, R^2, MSE, RMSE, and CVRMSE. The algorithms utilized were Extreme Gradient Boosting (XGB), Random Forest (RF), Bayesian Ridge (BR), Kernel Ridge (KR), CatBoost (CB), Light Gradient Boosting Machine (LGBM), Gradient Boosting (GB), Decision Tree (DT), Simple multi-layer perceptron artificial neural network (MLP).

7.5 Results

This chapter describes the outcomes of the experiments based on several evaluation metrics such as MAE, R^2, MSE, RMSE, and CVRMSE. The results are illustrated in Table 7.1.

Table 7.1. Algorithmic results per metric, for one step ahead load forecasting

Model	R^2 (0-1)	MAE (kWh)	MSE (kWh^2)	RMSE (kWh)	CVRMSE (0-1)
XGB	0.8538	0.1309	0.0328	0.1811	0.2911
LGBM	0.9332	0.0733	0.0150	0.1224	0.1968
CB	0.9183	0.0874	0.0183	0.1354	0.2177
KR	0.9259	0.0809	0.0166	0.1289	0.2072
BR	0.9256	0.0809	0.0167	0.1292	0.2077
GB	0.9318	0.0730	0.0153	0.1236	0.1988
RF	0.9029	0.1024	0.0218	0.1476	0.2373
DT	0.7475	0.1461	0.0566	0.2380	0.3825
MLP	0.9332	0.0785	0.0062	0.0785	0.1968

From the R^2 point of view, several models achieved high scores, greater than 0.9. More specifically, LGBM and MLP had the similar best R^2 (0.9332), with the GB being very close (0.9318). KR, BR, CB and RF also performed very well and above 0.9, having scores of 0.9259, 0.9256, 0.9183 and 0.9029 respectively. Finally, XGB and DT achieved lower R^2 scores, below 0.9 (0.8538 and 0.7475).

As far as MAE is concerned, GB and LGBM had almost identical errors (0.0730 and 0.0733 kWh), with MLP performing very close (0.0785 kWh). Furthermore, KR, BR, and CB had MAE lower than 0.1 (0.0809, 0.0809 and 0.0874, performing very well. Finally, RF, XGB, and DT had greater MAE than 0.1 (0.1024, 0.1309 and 0.1461), scoring the highest error.

Regarding MSE, RMSE and CVRMSE MLP was the most efficient model scoring 0.0062, 0.0785 and 0.1968 respectively. LGBM was the 2nd most accurate reaching 0.0150, 0.1224, and 0.1968 with GB almost similar scoring 0.0153, 0.1236, 0.1988. Moreover, KR (0.0166, 0.1289, 0.2072), BR (0.0167, 0.1292, 0.2077) and CB (0.0183, 0.1354, 0.2177) performed also very well. Finally, as in R^2 and MAE cases, RF, XGB, and DT were outperformed returning lower scores.

Overall, the findings revealed that LGBM, GB, and MLP were the most accurate models. More specifically, all three models were almost comparable, with the LGBM offering slightly better results for R^2, MLP significantly better scores for the squared error-based metrics (MSE, RMSE, CVRMSE) and the GB providing better results for MAE.

7.6 Conclusions

This research analyzes a case in time series forecasting by offering a unique technique for one step ahead of ELF that combines GDB technologies with ML and DL. It provides a clear guideline for developing correct insights for energy demands, but it also involves Neo4j for data integration, resulting in findings that can be matched towards other state-of-the-art approaches.

The results and the detailed comparison utilized indicated that the most accurate models were LGBM, GB, and MLP, with each one performing better for a specific metric (LGBM for R^2, GB for MAE and MLP for squared error-based metrics – MSE, RMSE, CVRMSE).

7.7   Limitations

The accompanying assertions are responsible for this study's restrictions. Information granularity and quality are two widely common issues in this research, as well as within the context of making accurate forecasts. Data should be complete, up-to-date, and widely available. Their quantity, while reliant on the kind of algorithm used, may represent a balancing challenge. In other words, whereas large data sets may be advantageous for model training, time and space complication might be a significant constraint. Model overfitting and outliers' detection are two extra-related issues.

7.7.1   Future work

Regarding possible future work, this work could be broadened by investigating the next positions:

(i)   Further expand the ELF utilizing several extra algorithms, such as LSTMs, SMVs etc.

(ii) Enhance the implemented algorithms by utilizing hyperparameter tuning via grid or random search.

(iii) Improve the suggested method by automating the procedure so that it may be used as a stand-alone option with only the necessary sets of input data.

(iv) Include ecological and financial Key Point Indicators (KPIs) in the prediction methodology, including power bills and carbon dioxide emissions. Following these KPIs, additional objects such as water bills, purchase records, soil or air contamination, and resource utilization could be included.

Ethical considerations

We would like to inform you that all information was gathered either open source or by subscription and is free to use.

Acknowledgements

We would like to thank CERTH for using their open-access database

Conflict of Interest

We do not have any conflicts of interest other than with staff working at the International Hellenic University.

References

[1] Brugnara, M., M. Lissandrini and Y. Velegrakis. 2022. Graph databases for smart cities. *IEEE Smart Cities*, University of Trento, Aalborg University.

[2] Mystakidis, A., N. Stasinos, A. Kousis, V. Sarlis, P. Koukaras, D. Rousidis, I. Kotsiopoulos and C. Tjortjis. 2021. Predicting covid-19 icu needs using deep learning, xgboost and random forest regression with the sliding window technique. *IEEE Smart Cities*, July 2021 Newsletter.

[3] Mystakidis, A. and C. Tjortjis .2020. Big Data Mining for Smart Cities: Predicting Traffic Congestion using Classification. 2020 11th International Conference on Information, Intelligence, Systems and Applications (IISA), pp. 1–8, doi: 10.1109/IISA50023.2020.9284399.

[4] Christantonis, K. and C. Tjortjis. 2019. Data Mining for Smart Cities: Predicting Electricity Consumption by Classification. 2019 10th International Conference on Information, Intelligence, Systems and Applications (IISA), pp. 1–7, doi: 10.1109/IISA.2019.8900731.

[5] Anagnostopoulos, T., A. Zaslavsky, K. Kolomvatsos, A. Medvedev, P. Amirian, J. Morley and S. Hadjiefthymiades. 2017. Challenges and opportunities of waste management in IoT-enabled smart cities: A survey. IEEE Transactions on Sustainable Computing, 2(3): 275–289, 1 July-Sept. 2017, doi: 10.1109/TSUSC.2017.2691049

[6] Rousidis, D., P. Koukaras and C. Tjortjis. 2020. Social media prediction: A literature review. *Multimedia Tools and Applications*, 79: 1–33. doi: 10.1007/s11042-019-08291-9.

[7] Perçuku, A., D. Minkovska and L. Stoyanova. 2018. Big data and time series use in short term load forecasting in power transmission system. *Procedia Computer Science*, 141: 167–174, ISSN 1877-0509, https://doi.org/10.1016/j.procs.2018.10.163.

[8] Huang, H., Z. Hong, H. Zhou, J. Wu and N. Jin .2020. Knowledge graph construction and application of power grid equipment. *Math. Probl. Eng.*, vol. 2020, no. January 2018, doi: 10.1155/2020/8269082

[9] Agrawal, S. and A. Patel. 2016. A Study On Graph Storage Database Of NoSQL. *International Journal on Soft Computing, Artificial Intelligence and Applications (IJSCAI)*, 5(1): 33–39.

[10] Patel, A. and J. Dharwa. 2017. Graph Data: The Next Frontier in Big Data Modeling for Various Domains. *Indian Journal of Science and Technology*, 10(21): 1–7.

[11] Ma, S., J.J. Li, C. Hu, X.X. Lin and J. Huai. 2016. Big graph search: Challenges and techniques. *Frontiers of Computer Science*, 10(3): 387–398.

[12] Singh, D.K. and R. Patgiri. 2016. Big Graph: Tools, Techniques, Issues, Challenges and Future Directions. *In:* 6th Int. Conf. on Advances in Computing and Information Technology (ACITY 2016), pp. 119–128. Chennai, India.

[13] Petkova, T. 2016. Why Graph Databases Make a Better Home for Interconnected Data than the Relational Databases. (2016). https://ontotext.com/graph-databases-interconected-data-relationaldatabases/

[14] Kaliyar, R. 2015. Graph Databases: A Survey. *In:* Int. Conf. on Computing, Communication and Automation (ICCCA2015), pp. 785–790.

[15] Robinson, I., J. Webber and E. Eifrem. 2015. Graph Databases New Opportunities for Connected Data, pp. 171–211.

[16] Zyglakis, L., S. Zikos, K. Kitsikoudis, A.D. Bintoudi, A.C. Tsolakis, D. Ioannidis and D. Tzovaras. 2020. Greek Smart House Nanogrid Dataset (Version 1.0.0) [Data set]. Zenodo. https://doi.org/10.5281/zenodo.4246525

[17] Gholamy, A., V. Kreinovich and O. Kosheleva. 2018. Why 70/30 or 80/20 relation between training and testing sets: a pedagogical explanation. Departmental Technical Reports (CS). 1209. https://scholarworks.utep.edu/cs techrep/1209.

[18] Ramos, M., A. McNabola, P. López-Jiménez and M. Pérez-Sánchez. 2020. Smart water management towards future water sustainable networks. *Water*, 12(1): 58. https://doi.org/10.3390/w12010058

[19] Toma, C., A. Alexandru, M. Popa and A. Zamfiroiu. 2019. IoT solution for smart cities' pollution monitoring and the security challenges. *Sensors*, 19(15): 3401. https://doi.org/10.3390/s19153401

[20] Lee, C., C. Lin and M. Chen. 2001. Sliding-Window Filtering: An Efficient Algorithm for Incremental Mining. *In:* Proceedings of the Tenth International Conference on Information and Knowledge Management. pp. 263–270.

[21] Alfares, H.K. and M. Nazeeruddin. 2002. Electric load forecasting: Literature survey and classification of methods. *Internat. J. Systems Sci.*, 33(1): 23–34.

[22] Salama, F., H. Rafat and M. El-Zawy. 2012. General-graph and inverse-graph. *Applied Mathematics*, 3(4): 346–349. doi: 10.4236/am.2012.34053.

[23] Vukotic, A. and N. Watt. 2015. Neo4j in Action, pp. 1–79. ISBN: 9781617290763.

[24] https://neo4j.com/blog/neo4j-3-0-massive-scale-developer-productivity/. Accessed: 2018.05.14.

[25] Kantarci, B., K.G. Carr and C.D. Pearsall. 2017. SONATA: Social Network Assisted Trustworthiness Assurance in Smart City Crowdsensing. *In:* The Internet of Things: Breakthroughs in Research and Practice. pp. 278–299. Hershey, PA, IGI Global.

[26] Rodriguez, J.A., F.J. Fernadez and P. Arboleya. 2018. Study of the Architecture of a Smart City. Proceedings, vol. 2, pp. 1–5.

[27] Bertot, J.C. and H. Choi. 2013. Big data and e-government: Issues, policies, and recommendations. *In:* Proceedings of the 14th Annual International Conference on Digital Government Research, pp. 1–10. ACM, New York.

[28] West, D.M. 2012. Big Data for Education: Data Mining, Data Analytics, and Web Dashboards. Gov. Stud. Brook. US Reuters.

[29] U.S. Department of Energy. 2019. Report on Smart Grid/Department of Energy. Available at https://www.energy.gov/oe/articles/2018-smart-grid-system-report, Retrieved Sep. 29.

[30] Christantonis, K., C. Tjortjis, A. Manos, D. Filippidou and E. Christelis. 2020. Smart cities data classification for electricity consumption & traffic prediction. *Automatics & Software Enginery*, 31(1): 49–69.

[31] International Energy Agency C. 2020. Electricity information overview (2020) URL: https://www.iea.org/reports/electricity-information-overview

[32] Alkhathami, M. 2015. Introduction to electric load forecasting methods. *J Adv Electr Comput Eng*, 2(1): 1–12.

[33] Koukaras, P., N. Bezas, P. Gkaidatzis, D. Ioannidis, D. Tzovaras and C. Tjortjis. 2021. Introducing a novel approach in one-step ahead energy load forecasting. *Sustainable Computing: Informatics and Systems*, 32: 100616.

[34] Ahmad, A., N. Javaid, A. Mateen, M. Awais and Z.A. Khan. 2019. Short-term load forecasting in smart grids: An intelligent modular approach. *Energies*, 12(1): 164, 10.3390/en12010164.

[35] Zhang, J., Y.M. Wei, D. Li, Z. Tan and J. Zhou. 2018. Short term electricity load forecasting using a hybrid model. *Energy*, 158: 774–781, 10.1016/j.energy.2018.06.012.

[36] Kuo, P.-H. and C.-J. Huang. 2018. A high precision artificial neural networks model for short-term energy load forecasting. *Energies*, 11(1): 213, 10.3390/en11010213.

[37] Raza, M.Q. and A. Khosravi. 2015. A review on artificial intelligence-based load demand forecasting techniques for smart grid and buildings. *Renew. Sustain. Energy Rev.*, 50: 1352–1372, 10.1016/j.rser.2015.04.065.

[38] Singh, A., K. Ibraheem, S. Khatoon, M. Muazzam and D.K. Chaturvedi. 2012. Load forecasting techniques and methodologies: A review. *In:* 2012 2nd International Conference on Power, Control and Embedded Systems. pp. 1–10.

[39] Fu, G. 2018. Deep belief network-based ensemble approach for cooling load forecasting of air-conditioning system. *Energy*, 148: 269–282.

[40] Amasyali, K. and N.M. El-Gohary. 2018. A review of data-driven building energy consumption prediction studies. *Renew. Sustain. Energy Rev.*, 81.

[41] Moon, J., Z. Shin, S. Rho and E. Hwang. 2021. A comparative analysis of tree-based models for day-ahead solar irradiance forecasting. 2021 International Conference on Platform Technology and Service (PlatCon), pp. 1–6, doi: 10.1109/PlatCon53246.2021.9680748.

[42] Khan, R.A., C.L. Dewangan, S.C. Srivastava and S. Chakrabarti. 2018. Short term load forecasting using SVM models. Power India Int. Conf., 8.

[43] Hao, H. and F. Magoules. 2012. A review on the prediction of building energy consumption. *Renew. Sustain. Energy Rev.*, 16(6): 3586–3592, 10.1016/j.rser.2012.02.049

[44] Lee, S.K. and S. Jin. 2006. Decision tree approaches for zero-inflated count data. *Journal of Applied Statistics*, 33(8): 853–865.

[45] Mystakidis, A., E. Ntozi, K. Afentoulis, P. Koukaras, P. Gkaidatzis, D. Ioannidis, C. Tjortjis and D. Tzovaras. 2023. Energy generation forecasting: Elevating performance with machine and deep learning. *Computing*, https://doi.org/10.1007/s00607-023-0164-y

[46] Mystakidis, A., E. Ntozi, K. Afentoulis, P. Koukaras, G. Giannopoulos, N. Bezas, P.A. Gkaidatzis, D. Ioannidis, C. Tjortjis and D. Tzovaras. 2022. One step ahead energy load forecasting: A multi-model approach utilizing machine and deep learning. *In:* 2022 57th International Universities Power Engineering Conference (UPEC), pp. 1–6, https://doi.org/10.1109/UPEC55022.2022.9917790

A Graph-Based Data Model for Digital Health Applications

Jero Schäfer and Lena Wiese

Institute for Computer Science, Goethe University, Frankfurt, Germany
e-mail: {jeschaef, lwiese}@cs.uni-frankfurt.de

The era of big data constantly introduces promising profits for the healthcare industry, but simultaneously raises challenging problems in data processing and management. The flood of automated systems in our smart cities generates large volumes of data to be integrated for valuable information which can also be applied to support in the healthcare context. Medical databases offer the capabilities to address these issues and integrate, manage and analyze the data for gaining deep insights. Especially modern graph databases which support the discovery of complex relationships in the ever-growing networks of heterogenous medical data. However, traditional relational databases are usually not capable of representing data networks in tables or revealing such relationships, but are still widely used as data management systems. This chapter discusses a methodology for transferring a relational to a graph database by mapping the relational schema to a graph schema. To this end, a relational schema graph is constructed for the relational database and transformed in multiple steps. The approach is demonstrated for the example of a graph-based medical information system using a dashboard on top of a Neo4j database system to visualize, explore and analyse the stored data.

8.1 Introduction

Smart Cities have the aim of improving living conditions for their citizens by means of a variety of digitized service offerings and a large-scale analysis of citizen-related data. A fundamental pillar for smart cities is the area of digital health due to the following reasons:

- Health and well-being of citizens are the major objectives of smart cities. Collecting the personalized health data of citizens is hence the foundation of analyzing living conditions or identifying adverse environments in cities.
- Health care is a main economic factor in cities and digital health can improve the timely delivery of health care services, as well as facilitate, optimize and modernize health care management procedures. In this sense, digital health can also relieve the burden involved with an aging society as well as shortages of qualified healthcare professionals.
- Availability of easy-to-use, secure and reliable digital health and telemedicine applications can improve the diagnosis and treatment of patients, in particular in regions where comprehensive health services are not regularly available or where patients have difficulties reaching the medical service center.
- Hospitals as a major unit in smart cities rely heavily on medical information systems in order to manage the every-day processes.
- Wearables (like smart watches) and smart phones are now-a-days widely used to collect health-related data on an individualized basis leading to a stark increase of health-related data, that can be applied to other aspects of smart cities (for example, analyzing smart mobility aspects by assessing the usage of bikes by individuals).

From a data management perspective, digital health systems provide the potential to integrate information and communication in healthcare more efficiently with the goal of sustainable benefits for all individuals.

8.1.1   Contributions

We propose a methodology for transforming a relational database schema into a graph data schema. The proposed method is general, intuitive and easy to apply. It constructs a schema graph from the relational schema and transforms it step-by-step into a graph model to be applied to a graph database management system. As an example, a medical application (and an implementation based on Neo4j and *NeoDash* – a Neo4j Dashboard Builder [5]) for storing and visualizing electronic health records, diagnostics and biomedical data is discussed. This example illustrates the different transformation steps of the schema graph from the relational to the graph database schema in a real-world application.

8.1.2   Outline of the chapter

Before describing the proposed schema transformation, an overview of related work in the field of (bio)medical graph data management and a motivation for the application of a graph database system are given in Section 2. Then, Section 3 outlines the formal description of our proposed method in this chapter. This includes the transformation of a relational database into a graph database via the construction of a schema graph from the relational data model (Section 3.1) and its transformation towards a graph data model (Section 3.3). The descriptions of the schema graph construction and transformation are accompanied by small examples for illustrative purposes (Sections 3.2 and 3.4). The application of the proposed technique for schema transformation is also analyzed using a real-world, medical dataset (Sections 4.1 and 4.2). The feasibility of the approach is therefore investigated in Section 4.3 under the demonstration of a web application-based medical information system for the visual exploration and analysis of the underlying Neo4j graph database system using a simple, yet effective dashboard built with NeoDash.

8.2 Background

We can define digital health to span a variety of applications and information systems that collect individualized person-related data on a large scale. Those information systems need a flexible data management solution in the backend that can incorporate different kinds of input data (formats) and at the same time provide an easily extensible data model that also allows for a user-friendly visualisation of information. Moreover, an optimized (graph-based) data structure can build a reliable and efficient backbone for an AI-based evaluation of large digital health data sets.

As compared to tabular data storage, graph structures may often be better suited for medical data management. Looking in more detail at biomedical data one can observe that many of these data are inherently graph-structured; an overview of graph databases in the biomedical domain is provided in [10]. These graph structures manifest at the low level of biochemistry as well as the high level of disease and patient management:

- For example – on the low level of biochemical processes – gene regulation can be expressed in a network where different components interact: proteins interact with other proteins [8], or genes interact with genes [9] as well as for analysis and visualisation of dynamic gene regulatory networks [12].
- On the high level of electronic health records, the data of individual patients can be linked via their symptoms to other patients with similar symptoms or to supporting documentation on similar diseases [6].
- Unified vocabularies (thesauri, ontologies, taxonomies), that are also graph-shaped, have to be involved to enable matchings beyond simple string equality: semantic similarity between terms (for example, disease categories like "ulna fracture" and "broken arm") will produce more relevant and accurate cohorts [2].
- Moreover, during the run of this project we fuse different notions of similarity for patients (different similarities for different data types) into one global similarity in a so-called similarity fusion network [11], which is again a graph structure.
- Another graph approach is the one of disease networks [1, 3]; by observing deviations of a disease network from a healthy state, diseases and their causes can be identified.

What can be seen from these examples is that it is advantageous to combine medical data in a graph to achieve a full-scale personalised medical treatment: several different data sources have to be combined to obtain a comprehensive view of a patient's health – hence requiring a graph structure with lots of interconnections between several data sets.

8.3 Relational schema transformation

We present a technique to transform and migrate a relational database into a Neo4j property-graph database to be able to represent former relational data in a knowledge graph for more sophisticated analyses of complex relationships. The transformation is achieved by constructing a schema graph **RG** from the relational database schema $\mathbf{R} = \{R_1(X_1), \ldots, R_z(X_z)\}$. This schema graph is then subsequently mapped with multiple transformation steps to the schema of the graph database. The initial schema graph $\mathbf{RG} = \langle N, E \rangle$ consists of nodes $n \in N$ created for each attribute $a \in X_i$ of the former schemata $R_i(X_i)$ of the relations R_i in **R** and edges $e \in E$ modeling the primary (PK) and foreign key (FK) dependencies between the attributes and relations.

The schema graph **RG** is modified by merging, replacing or removing nodes and edges of **RG** that satisfy specific properties and conditions to compress the attributes and their dependencies into connected, logical entities modeling the data graph. This way the former foreign key dependencies can be intuitively resolved as connections between the logical entities instead of additional join tables modeling the one-to-many or many-to-many relationships. The converted schema graph **RG** then consists of entities (nodes N) and their relationships (edges E) according to which instances can be stored and queried by the corresponding graph database system.

The required transformation steps, that are explained in more detail below, can be summarized as follows:

1. Create nodes labeled with "$R_i.a$" for each attribute a of any relation schema $R_i(X_i)$.
2. Merge all the nodes of composite PK attributes of the same relation into a single PK node labeled "$R_i.PK$".
3. Create directed edges (n_p, n_a) from PK nodes n_p to nodes n_a of other attributes a of the same relation R_i.
4. Create directed edges (n_f, n_p) from FK nodes n_f to the respective PK nodes n_p.
5. Merge sinks n_s (i.e. nodes without any outgoing edges) that are all connected to the same PK node n_p and have only one incoming edge (n_p, n_s) into one node labeled "$R_i.attributes$".
6. Merge PK hubs n_h (i.e. nodes with incoming and outgoing edges) labeled $R_i.a$ with only one outgoing edge (n_h, n_s) to a (merged) sink n_s with this sink. These new entity nodes n_E are labeled "R_i" and contain all attributes from the previously merged hub and sink.

7. Replace sources n_s (i.e. nodes without incoming edges), that are connected to exactly two entity nodes n_{E_1} and n_{E_2}, and their two edges (n_s, n_{E_1}) and (n_s, n_{E_2}) by an undirected edge $e = \{n_{E_1}, n_{E_2}\}$ connecting n_{E_1} and n_{E_2} directly.

- Case 1: If n_s has no other edges, no other actions are required.
- Case 2: If n_s has an edge to a (merged) sink n_q, add the attribute(s) represented by n_q as property to e and remove n_q, too.
- Case 3: If n_s has an edge to a hub n_h with only one other edge to an entity node n_E containing only one additional attribute a next to the identifying attribute(s), add a as property to e and remove n_h and n_E from the schema graph.

If none of the above cases is applicable, no merge is performed but the directed edges (n_s, n_E) to any entity node n_E are transformed into undirected ones.

8. Resolve FK relations by edges:
- Case 1: Replace FK relations indicated by hubs n_h with one incoming edge (n_{E_1}, n_h) from an entity node n_{E_1} and one outgoing edge (n_h, n_{E_2}) to an entity node n_{E_2} by an undirected edge $\{n_E, n_{E2}\}$.
- Case 2: If the FK relation is a source n_s labeled "$R_i.a$" with outgoing edges to an entity node n_E and to a (merged) sink n_q, first merge n_s and n_q (with all attributes except the FK attribute) into an entity node $n_{E'}$ labeled "R_i". Then, connect the entity nodes n_E and $n_{E'}$ with an undirected edge $n_E, n_{E'}$.

9. Transform any node n, which is not an entity yet, into an entity node and each directed edge (n_a, n_b) into an undirected edge $\{n_a, n_b\}$ between the same nodes n_a and n_b.

8.3.1 Schema graph construction

Given a relational database schema $\mathbf{R} = \{R_1(X_1), \ldots, R_z(X_z)\}$, we construct a node n_a for each of the attributes $a \in X_i$ of the relation schemata $R_i(X_i) \in \mathbf{R}$. Each node n_a is then labeled with "$R_i.a$" where R_i is the name of the corresponding relation and a the name of the attribute. For each relation schema $R_i(X_i) \in \mathbf{R}$ with a composite PK, one single node labeled "$R_i.PK$" is created for convenience and it merges all the composite PK attributes of individual nodes. By doing so, the complexity of the schema graph $\mathbf{RG}$ is reduced due to the decreased number of nodes and edges and the subsequent transformations become more intuitive. Otherwise, there would be additional edges between the attribute nodes that compose the PK of a relation R_i and edges from each of the PK attribute nodes to the non-key attributes of this relation or to key attributes of other relations denoting FKs.

After the nodes are created and composite PK attribute nodes have been merged, the edges denoting the relationships are introduced. From each PK node n_a of relation schema $R_i(X_i) \in \mathbf{R}$ a directed edge $e = (n_a, n_b)$ is established for each of the non-key attributes $b \in X_i$ of the same relation. Similarly, each node n_a of relation schema $R_i(X_i) \in \mathbf{R}$ for a single FK attribute a is connected with an edge $e = (n_a, n_b)$ to the corresponding PK node n_b of $R_j(X_j)$. It should be noted here that we do not consider composite FKs here as they can be handled analogously to the composite PKs by collapsing the respective attribute nodes into a single FK node. The introduced edges mirror the logical connections between the entities and are formed based on the join conditions.

8.3.2   Schema graph construction example

Consider the four relations A, B, C and D as shown in Figure 8.1 (a). The primary key attributes of the relations are underlined and the arrows indicate FK dependencies. Following the construction steps explained before, for each relation $R_i \in \{A, B, C, D\}$ create a node for each of its attributes, e.g., three nodes labeled "$A.a$" "$A.a_1$" and "$A.a_2$" are created for relation A (Figure 8.1 (b)). Then, the nodes representing composite PKs are aggregated into PK nodes. Here, the PK $(C.a, C.d)$ of relation C (i.e., an associative relation modeling the one- or many-to-many relationship between A and D) is transformed into one node labeled "$C.PK$". The edges between PK nodes and the non-key attributes of the same relation R_i are established, e.g., two arrows point from the node labeled "$B.b$" to the nodes labeled "$B.a$" and "$B.b_1$". Finally, the FK dependencies connect FK attribute nodes to the corresponding PK nodes of the other relations, e.g., the FK dependencies of relation C connect the PK node of C to the PK nodes of A and D, respectively.

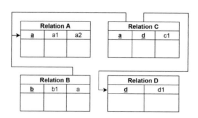

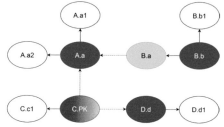

| (a) Relations with PKs denoted by underlines and arrows indicating FK dependencies. | (b) Constructed schema graph **RG** for the relations in (a) with PK nodes (purple), FK attribute nodes (yellow) and FK relationships (dashed arrows). |

Figure 8.1. Schema graph construction example.

8.3.3   Schema graph transformation

Furthermore, the schema graph **RG** is compressed by merging the sinks $n_s \in N$ (i.e. nodes without any outgoing edges) that are all connected to the same PK node n_p and only have one incoming edge (n_p, n_s). The respective sink nodes n_s are then merged into one node labeled with "$R_i.\text{attributes}$". This aggregation of the non-key attribute nodes is the first step to modeling entities in the design of the final graph database schema obtained from **RG**. To create the first entities, all hubs $n_h \in N$ of some attribute $a \in X_i$, that are PK nodes of R_i with only one outgoing edge (n_h, n_s) to a (merged) sink n_s representing attribute(s) $b \in X_j$, are merged with the corresponding sink n_s into one new node n labeled R_i. The new node n then has the set of attributes $\{a, b\}$ (or $\{a\} \cup b$ if b denotes a set of attributes from a merged sink n_s). This ensured that the constructed node n acts as an entity modeling the relation R_i together with the entity-specific attributes and, hence, the graph data model becomes even more intuitive and less inflated.

The next step is to resolve associative relations R_w that model the one- or many-to-many relationships between two other relations R_i and R_j. To this end, source nodes n_s with two outgoing edges $e_1 = (n_s, n_{E_1})$ and $e_2 = (n_s, n_{E_2})$ to two entity nodes n_{E_1} and n_{E_2} are processed. These nodes make up the fundamental units, i.e. the real-world entities, and have a label and attributes. The node n_s and the edges e_1 and e_2 are removed from **RG** and all the information (attributes) of n_s is incorporated into an undirected edge $e = \{n_{E_1}, n_{E_2}\}$ between n_{E_1} and n_{E_2}. If e_1 and e_2 were the only two edges attached to n_s, no further processing is required (case 7.1). In the case of one additional outgoing edge (n_s, n_q) to a (merged) sink n_q, the attribute set represented by n_q is added as the attribute set of the new undirected edge e (case 7.2). Also, if n_s has an outgoing edge (n_s, n_h) to a hub node n_h with only one other edge (n_h, n_E) to an entity node n_E where n_E only contains a single additional attribute b next to the identifying attribute(s) a, add b as property to e and remove n_h and n_E from **RG** (case 7.3). However, if there are more edges originating from n_s than described above or if the additionally connected hub n_h from the latter case does not match the constraints, i.e., none of the cases apply, no replacement of n_s, e_1 and e_2 is possible. In this case n_s becomes an entity as well and the edges e_1 and e_2 are made undirected.

As a further processing step towards the final graph data model, the remaining FK relations in the notion of the relational schema were also transformed into edges modeling relationships between the corresponding entities. The FK relations were either indicated by hubs n_h with exactly one incoming edge (n_{E_1}, n_h) from an entity node n_{E_1} and one outgoing edge (n_h, n_{E_2}) to an entity node n_{E_2} or sources n_s linking via (n_s, n_E) to an entity node n_E and (n_s, n_q) to a (merged) sink n_q. In the first case, such a hub n_h (and its edges) are simply replaced by an undirected edge $e = \{n_{E_1}, n_{E_2}\}$ directly connecting n_{E_1} and n_{E_2}. The sources n_s (case 8.2) are treated differently as before establishing an undirected edge between the entity node n_E and the next entity node $n_{E'}$ a second entity node has to be created first. This entity is drawn from merging the source n_s with the (merged) sink n_q into an entity node $n_{E'}$ with the attributes from n_s and n_q. The foreign identifier from the attributes of n_s is omitted in the context of this merge as the new edge incorporates this relation inherently. With the new entity node $n_{E'}$, the relationship in terms of an edge $\{n_E, n_{E'}\}$ between the nodes n_E and $n_{E'}$ is established.

To finalize the graph data model, the remaining non-entity nodes are transformed into entity nodes as the relationships can not be resolved further. Any directed edge (n_a, n_b) in **RG** between any two nodes n_a and n_b is replaced by an undirected edge n_a, n_b. Then, the final graph data model is captured by the fully transformed schema graph **RG**, that now consists of only relationships (edges) and entities (nodes) with properties and labels.

8.3.4. Schema graph transformation examples

Entity creation and FK relationships

Referring again to the four toy relations from Section 3.2, the next transformation of the graph is obtained by aggregating non-key attributes through merging sinks of **RG** only connected to the same PK node and afterwards creating entity nodes from hubs and (merged) sinks. Figure 8.2 (a) depicts the graph with the three entities *A, B* and *D* (magenta) that are obtained in this way. Furthermore, the FK dependencies between the three entities are resolved. The PK node "*C.PK*" is replaced by an undirected edge that also absorbs the attribute $C.c_1$ as this is a sink connected to the replaced PK node (case 7.2). The hub labeled "*B.a*" is also replaced by an undirected edge connecting entities *A* and *B* (case 8.1) and the final schema is shown in Figure 8.2 (b).

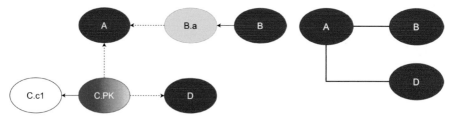

(a) Transformed schema graph with entity nodes (magenta), PK nodes (purple), FK nodes (yellow) and FK relationships (dashed arrows).

(b) Final schema graph consisting of connected entities *A, B* and *D*.

Figure 8.2. Schema graph transformation example.

8.3.5 Transformation of *n*-ary associative relations

Consider the four relations *R, S, T* and *J* as shown in Figure 8.3 (a) depicting a typical associative relation *J* joining the other three relations. This ternary relationship can not be modeled by an edge in the schema graph **RG**. After the proposed transformation steps are applied, the schema graph with three entities and an interconnecting source node labeled "*J. PK*" is obtained (Figure 8.3 (b)). This source node then is transformed into an entity in the last step of transformation where also the three directed edges become undirected relationships. This example shows how *n*-ary (*n* > 2) relationships are modeled by a star-like schema, i.e., the associative relation becomes an entity connecting in the middle of all the other relations.

Binary associative relations with supplementary relations

In contrast to this, consider that the attribute *J.t* of relation *J* from the previous example is not part of the PK of relation *J*, but only a FK attribute referring to relation *T*. In this case, the relationship between *R* and *S* is characterised by

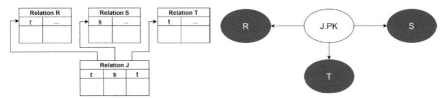

(a) Ternary associative relationship with PKs denoted by underlines and arrows indicating FKs.

(b) Transformed schema graph for the ternary associative relationship between *R*, *S* and *T*.

Figure 8.3. Schema graph transformation counter example *n*-ary associative relationship.

information stored in the supplementary relation *T*. This scenario leads to a transformed schema graph **RG**, where the PK node "*J.PK*" connects to the two entity nodes labeled "*R*" and "*S*" and to the entity node labeled "*T* " via the FK attribute node labeled "*J.t*". The additional information modeled by relation *T* can be absorbed directly into the relationship when introducing an undirected edge between *R* and *S* that also incorporates the attributes of *T* (case 7.3). The final graph data model then consists of two entities that are connected by an undirected edge with additional attributes characterising the relationship according to *T*.

8.4 Graph-based medical information systems

8.4.1 Application of the schema transformation

This section discusses the application of the proposed relational to graph data schema transformation to a real-world, medical application. The considered relational database with medical data consists of routine patient health records. It contains, amongst other cases, a collection of cases of pediatric acute lymphoblastic leukemia (ALL) patients. The patient data are related to various other information, such as diagnostics and analytics. One of the main components are biological samples obtained from patients that represent complex biomedical data extracted with novel technologies, e.g., *next generation sequencing* (NGS). The findings of the processing of data about gene fusions and mutations reveal valuable insights and are the main drivers for the stratification of leukemia patients [4]. The relational schema is shown in Figure 8.4.

The Patient relation is one of the central entities of the database schema and stores patients' personal information, such as name, gender or date of birth (dob). It is connected to multiple associative relations, namely MaterialPatient, ProjectPatient, FamilyPatient, DiagnosisPatient and OrderPatient, which all relate the patients to additional data, like the samples taken from them (Material), e.g., blood or cell samples, or the research projects they participate in (Project). The Analysis relation adds the biomedical component of different types of analyses like arrayCGH for detecting loss or gain of genomic regions. The analyses, that are ordered for a certain patient for

Figure 8.4. Relational schema of the medical database. The tables represent the relations (PK bold) with attribute names and types. The lines between the relations indicate FK dependencies.

diagnostic purposes, for instance, relate the patients and their materials to the important, decisive analytical results, which in turn can be used by doctors for personalizing the treatment or researchers to gain knowledge.

The application of the proposed schema transformation approach leads to a first relational schema graph **RG** = ⟨*N, E*⟩ with (merged) attribute nodes, PK nodes, and PK-to-attribute & FK-to-PK edges. It is obtained by applying the transformation steps 1-5. For the sake of better visualisation, the schema graph is split into two parts **RG₁** ∪ **RG₂** (Figures 8.5 (a) and 8.5 (b)). The PK nodes of the graph are visualised with a thick outline and source and sink nodes are colored green and red, respectively. The arrows between the nodes indicate the directed edges for PK-attribute or FK-PK connections according to the schema graph construction rules. The first partition **RG₁** shows the schema graph of the `Patient, Family, Project, Diagnosis` and `DiagnosisAddition` relation, that all connect via associative relations. The second partition **RG₂** consists of the other relations involving the patient material and the analytic process. Both partitions **RG₁** and **RG₂** share the PK node of the Patient relation labeled "*Patient.id*", which would be the central node in a non-partitioned illustration, and the corresponding attribute node "*Patient.attributes*".

The further transformations of the nodes *N* and edges *E* towards the final graph data schema create the first entities merging the PK hub nodes (step 6) with the corresponding (merged) attribute nodes. This leads to the entities, which aggregate the entity-specific properties for patients, projects, families, diagnoses

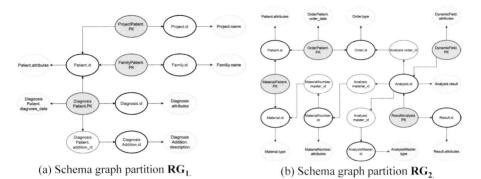

(a) Schema graph partition **RG₁**.   (b) Schema graph partition **RG₂**.

Figure 8.5. Partitioned schema graph **RG** = **RG₁** ∪ **RG₂** with PK nodes (thick outline), sink nodes (green) and source nodes (red) highlighted.

(additions), orders, analyses (masters), results and materials as shown in Figure 8.6 (left). Then, the source nodes connecting to two entity nodes are replaced by a direct edge between the entity nodes to model the relationship in the graph data schema, i.e., the relationships "*InProject*" and "*InFamily*" are added. It is possible according to transformation step 7.3 to replace the source node labeled "*DiagnosisPatient. PK*" as well as the attribute node and the FK dependency to the "*DiagnosisAddition*" entity by an edge labeled "*HasDiagnosis*". This edge also has the attribute `diagnosis_date` of the former `DiagnosisPatient` table and the attribute `DiagnosisAddition.description` as this is the only non-PK attribute and, thus, can be compressed into the new relationship. This leads to the final schema graph partition **RG₁** as depicted on the right side of Figure 8.6.

The same procedure as for **RG₁** is also done in parallel for the partition **RG₂**. Originating from the version shown in Figure 8.5 (b), the entities are again generated by merging hubs and their (merged) attribute nodes. This yields the entities of patients (same entity as in **RG₂**), orders, analyses (master), results and

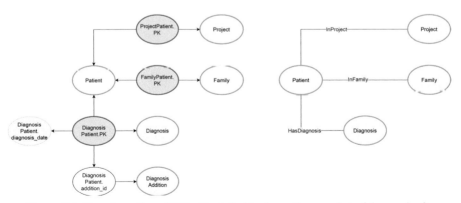

Figure 8.6. Transformation of **RG₁**: The left side shows the snapshot of the graph after creating entity nodes by merging hubs. The right side displays **RG₁** after additionally applying the transformation steps 7-9.

materials (numbers) (left side in Figure 8.7). For these entities, relationships are established by replacing the sources labeled "*OrderPatient. PK*", "*MaterialPatient. PK*" and "*ResultAnalysis. PK*" by edges (steps 7.1 and 7.2) labeled "*HasOrder*", "*HasMaterial*" and "*HasResult*", respectively. After introducing these edges, the updated schema graph partition **RG₂** (right side in Figure 8.7) still contains several FK dependencies and non-entity nodes, e.g., the FK attribute node "*Analysis. order id*" connecting the performed analyses and their requests, that are about to be resolved.

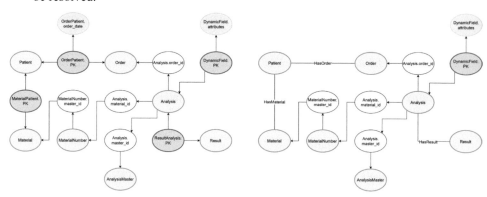

Figure 8.7. Transformation of **RG₂**: The left side shows the creation of the entities according to transformation step 6. The result of replacing hubs by edges to connect entities (step 7) is visualised on the right side.

The remaining FK dependencies indicated by hubs, that are FK attribute nodes between two entities, are replaced by edges modeling the relationships between the entities (step 8). In our use-case this leads to several new edges in **RG₂**. For example, the edge labeled "*OnMaterial*" links any execution of an analysis (e.g., RNA sequencing) to a biomedical sample (e.g., RNA) taken from a patient. In this context, the DynamicField relation, which is represented by the source labeled "*DynamicField.PK*" and the sink labeled "*DynamicField. attributes*", is also modeled by an entity as the source and sink are merged and the FK dependency on Analysis.id is represented by an undirected edge between

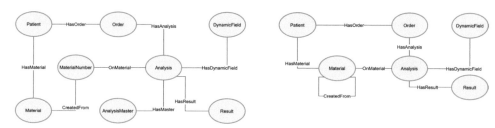

Figure 8.8. Transformation of **RG₂** continued: The final schema after the application of the last transformations (steps 8 and 9) is illustrated on the left side. The right side depicts some manual, use case-specific adaptations of the final schema.

the new entity and the analysis entity. Hence, the second partition of **RG** is composed as demonstrated by the graph on the left side of Figure 8.8.

8.4.2 Manual schema fine-tuning

Based on the given use-case, the schema graph, as defined in the partition RG_2, can be refined through some other modifications. This results in a more intuitive design of our graph database schema. As the `AnalysisMaster` relation only comprises the `id` and `description` attributes, the information provided to each of the analysis instances can be directly incorporated into the entity, too. Therefore, an additional attribute named *description* could be added to the attribute set of the analysis entity, which replaces the analysis master entity and the reference from the analysis entity to it completely. However, as the number of analysis types, i.e., the number of instances of the "*AnalysisMaster*" entity, is relatively small, it is more efficient to specify sublabels for each of the different analysis types under the shared general "*Analysis*" label. This way the graph can be traversed more intuitively when querying for a specific type of analysis or when querying for any type of analysis. The modification of the graph is shown in the right side of Figure 8.8, where the analysis master entity was omitted and instead incorporated as a label hierarchy. The analysis entity represents multiple entities (not visualized here) that have the common superlabel "*Analysis*" and a more specific sublabel based on the analysis type, e.g., "RNASeq".

A second modification is achieved by restructuring the hierarchy of materials, i.e., the biomedical samples taken from a patient are usually cultivated or prepared in any form resulting in a chain of materials, where each stage originates from the predecessor. Hence, the hierarchy of the materials is mapped by a self-reference of the entity "*Material*". This recursive relationship provides an intuitive way of modeling the hierarchy with main materials and sub-materials, which are produced from the main materials. The entity labeled "*MaterialNumber*" is merged into the "*Material*" entity and the latter entity is then directly related to an analysis via the relationship "*OnMaterial*" (right side of Figure 8.8).

8.4.3 Web application-based medical information system

The final graph database model is then composed from joining the two schema graph partitions RG_1 and RG_2 at the shared patient entity with the fine-tuning as described previously. In the backend, a Neo4j database system employs the derived graph schema **RG** as data model to store the data obtained from the relational database. The relevant data, that are fed into the database, are extracted from the relational database using SQL queries. In particular, a set of Python scripts is used to obtain the cohort of ALL-diagnosed patients by issuing the corresponding SQL queries that first retrieve the personal information of the affected patients like id, name, sex or date of birth. Then, the data related to the patients according to the graph schema **RG** are queried successively, e.g., biological samples (material) or analytical results (analysis sublabels).

Object graph mapping

The extracted data are mapped from the SQL query results to the corresponding nodes and edges in the graph database via Neomodel [7]. Neomodel is a lightweight tool for mapping objects to graph structures for the Neo4j database system. It provides an easily usable API that allows to define the graph model programmatically and to transfer the constructed objects to the graph in the Neo4j database. The patient cohort is then mapped to interconnected nodes in the graph structure as defined by RG by instantiating objects and inter-object relationships according to the extracted facts.

Dashboard builder

A dashboard is then built with NeoDash, a dashboard builder for Neo4j databases. This tool uses straightforward Cypher queries entered in a built-in editor to extract information from the data graph for visualization in different charts, e.g., bar or graph charts. The dashboard itself is composed of two main components, pages and reports. Each dashboard contains one or more pages that enable a conceptual separation of contents. The separation supports different perspectives of the same data. Each of the pages contains multiple reports that render a chunk of information obtained as the result of a single Cypher query against the database. For example, the report can show a distribution over the data in a bar chart or plot a subgraph of the data. The dashboard implemented in a NeoDash instance is deployed as a user-friendly web application.

We implemented three pages in our dashboard that visualise information on different levels of granularity. One page deals with the full cohort of pediatric ALL cases and gives a summarising overview of the cohort and its statistics. The second page concentrates on the analysis of a subgroup of patients based on the selection of a certain genetic fusion and allows the user to explore the subgroups relationships and for leukemia research relevant information like karyotype or aneuplouidy. The last page visualises the characteristics and healthcare data for an individual target patient and suggests similar patients. In the following subsections, we demonstrate the dashboard pages as implemented in the public demo version of Graph4Med[1]. The demo tool uses random and synthetic patient data generated with Synthea[2] for the sake of privacy.

Cohort page

This dashboard page yields the user a broad overview of the cohort of the extracted ALL patients (Figure 8.9). The first information is the size of the demo cohort given by the number at the top left report that is fed by a simple query returning the count of patient nodes. The page comprises a report rendering the distribution of the age of the patients in a bar plot which is additionally grouped by patients'

---

[1] http://graph4med.cs.uni-frankfurt.de/
[2] https://synthetichealth.github.io/synthea/

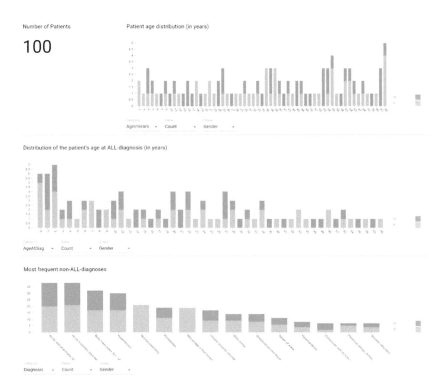

Figure 8.9. Cohort page with summarizing statistics on the patient cohort: The cohort size (top left), age distribution (top right), age at ALL-diagnosis distribution (middle) and mostfrequent non-ALL diagnoses (bottom). The colors of the bars indicate male (M, orange) and female (F, green) patients.

gender (top right). The grouping is indicated by the stacked colors of each of the bars and the gender attribute is only binary for the sake of simplicity. The second bar plot in the middle of the page represents the gender-grouped distribution of the age at which the ALL diagnosis was made. The last bar plot on the page's bottom illustrates the gender-grouped distribution of the most frequent diagnoses next to ALL throughout the cohort, e.g., AML (Acute Myeloid Leukemia) or (non)-pediatric MDS (Myelodysplastic Syndrome).

As the Cypher queries are written down in the built-in query editor, it is possible with only minor effort to extend these queries, such that the plots can be interactively changed by the user. For example, the user could then choose whether to display the absolute or relative frequencies for each of the age values or alternative groupings could be seamlessly applied and displayed, e.g., grouping by minimal residual disease instead of gender.

```
MATCH (n:Patient)
WITH n, duration.between(date(n.dob), date()) AS age
RETURN  age.years AS AgeInYears,
```

```
        n.gender AS Gender,
        COUNT (DISTINCT n)  AS Count
ORDER BY age.years
```

This Cypher query populates the "Patient age distribution" report (Figure 8.9) with patient counts grouped implicitly by age and gender. The dashboard can then simply render the bar chart with the data returned from the query. For example, the relative frequency can be chosen as an alternative for the absolute frequency by the user when adding

```
MATCH (:Patient)
WITH COUNT(*) AS cnt
...
RETURN ..., 1.0 * COUNT(DISTINCT n)/cnt AS Frequency
ORDER BY age.years
```

to the above query. The equivalent SQL query to the unmodified version is very similar for this use case:

```
SELECT  datediff(year, p.dob, getdate()) AS AgeInYears,
        p.gender AS Gender,
        COUNT(DISTINCT p.id) AS Count
FROM Patient p
GROUP BY datediff(year, p.dob, getdate()),
         p.gender
ORDER BY AgeInYears
```

In contrast to the Cypher query, the grouping for the age and gender of the patients has to be declared explicitly in SQL. The explicit grouping and the double function call of `datediff(year, p.dob, getdate())` for computing the age in years in SELECT and GROUP BY make the SQL query less comprehensible.

Subgroup page

The dashboard page for analysing subgroups of patients is shown in Figure 8.10. The top row of reports gives an overview over the whole cohort by statistics on the amount of detected fusions per patient (top left) and the ten most frequent detected fusions plus the frequency of occurrence of hyper-/hypodiploidy (top mid). To ease the fusion selection (mid left) by the user, the top right report gives a table of fusions and their alternative names. Upon the selection of a fusion, the affected reports are reloaded and the Cypher queries are re-evaluated under the instantiation of the included variable with the selected fusion value. For the subgroup of patients, i.e., those that were detected the selected fusion, age distribution plots (center and mid right), an interactive subgraph revealing the relationships between patients and fusions (bottom left) as well as a tabular overview with analytical information (bottom right), e.g., chromosomes or aneuploidy type, are contained in the dashboard page.

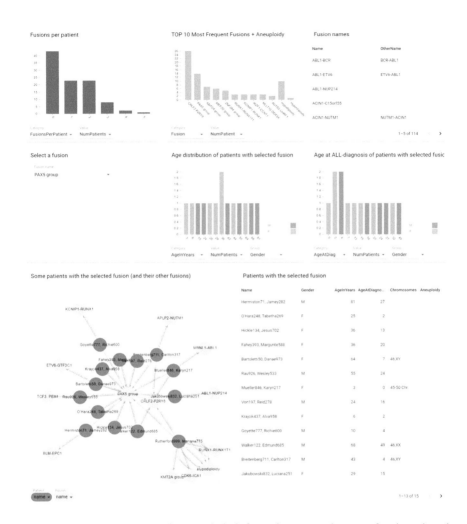

Figure 8.10. Subgroup page about analytic information on a subgroup of patients based on a selected fusion. The top row shows summarising statistics for the whole cohort and a table with (groups of) fusions and their other names. Age distributions (middle row) as well as a graph-based (bottom left) and tabular overview (bottom right) are shown for the subgroup of patients with the selected fusion (middle left).

```
MATCH (p:Patient)-[:HasOrder]-(:Order)-[:HasAnalysis]
     -(:Analysis)-[:HasFusion]-(f:Fusion),
     (p)-[d:HasDiagnosis]-(diag:Diagnosis)
WHERE f.name = neodash_fusion_name
  AND diag.name = "ALL"
WITH p, f, d, diag
OPTIONAL MATCH (p)-[:HasOrder]-(:Order)-[:HasAnalysis]-
               (a2:ArrayCGHAnalysis)
```

```
OPTIONAL MATCH  (a2)-[:HasAneuploidy]-(y:Aneuploidy)
OPTIONAL MATCH  (p)-[:HasOrder]-(:Order)-[:HasAnalysis]-
                (a3:KaryotypeAnalysis)
RETURN ...
```

The Cypher query above, fetches the data for the tabular report summarising the analytical information for the patients of the subgroup (leaving out the projection for the sake of simplicity). The variable neodash_fusion_name is the parameter for the selection of a fusion by the user and is replaced by the selected fusion name via the selection report shown at the left of the middle row of Figure 8.10.

```
SELECT ...
FROM   Patient p, OrderPatient, Order o, Analysis a,
       DynamicField df, DiagnosisPatient dp, Diagnosis diag
LEFT JOIN OrderPatient op2 ON p.id = op2.patient_id
LEFT JOIN Analysis a2 ON op2.order_id = a2.order_id
... joins aneuploidy/karyotype information ...
WHERE ... join conditions ...
      AND df.x = neodash_fusion_name
      AND diag.name = "ALL"
      AND ... selection analysis types ...
```

The sketched SQL query to obtain the same results as the previous Cypher query is relatively long, due to the explicit listing of the join conditions of all the involved tables. The join conditions to link the relations were omitted here for better readability. Due to the associative relations like OrderPatient or DiagnosisPatient, the joining conditions inflate the query structure massively. This example query demonstrates the strength of the Cypher query language with the much more intuitive and concise path notation, instead of the lengthy SQL statements involving multiple associative relations and the corresponding join conditions.

The modeling of the analysis subtypes by the introduced sublabels in the graph data model also makes it more convenient to formulate queries where it is directly visible whether the query targets single or multiple analysis types, e.g., (p)-[:HasOrder]-(:Order)-[:HasAnalysis]-(a2:ArrayCGHAnalysis). The SQL query, in contrast, needs to include (based on the relational schema) selections for the corresponding analysis type for each of the joined Analysis table instances (also omitted here).

Patient page

The last page of the Graph4Med dashboard enables the user to navigate the individual case of a selected patient. From investigations on a subgroup of patients with a certain fusion, the user might find interesting cases and explore them separately in this dashboard page. The first report at the top left as shown in Figure 8.11 performs the selection of a patient by id. With the selection of an individual, the dashboard shows a tabular overview of the patients diagnostic information in the form of the executed analyses and their results (top right). A subgraph comprising the patient's data in a comprehensive and interactive report is also generated (bottom left).

For further exploration, the similarity between the target patient and all other patients is calculated and the most similar ones are shown in a graph (bottom right). The similarity is computed as the Jaccard similarity

$$J(A, B) = \frac{A \cap B}{A \cup B}$$

between the set of fusions, aneuploidy type and diagnoses of the target patient and the other patients. The color and thickness of the "SimilarTo" relationships pointing to the other patients mirror the similarity, i.e., thick and dark green arrows correspond to a high similarity whereas thin and light green ones refer to a smaller degree of similarity.

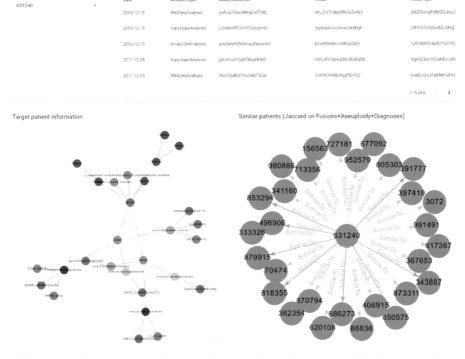

Figure 8.11. Patient dashboard page displaying information about the individual case of a single patient. Upon selection (top left), a table with analysis data (top right) and a graph with various data related to the target patient (bottom left), e.g., materials, analyses and diagnoses, are created. Similar patients are linked in the graph on the bottom right, where color and thickness of the "SimilarTo" relationship indicate the degree of similarity.

```
MATCH (p:Patient)
WHERE p.patient_id = target_patient_id
OPTIONAL MATCH (p)-[ho:HasOrder]-(o)-[ha:HasAnalysis]-(a)
OPTIONAL MATCH (a)-[hr:HasResult]-(r:Result)
...
RETURN *
```

This sketched Cypher query is used to construct the subgraph of a selected patient (with id `target_patient_id`) comprising the related information to the patient, e.g., analyses and detected fusions.

```
SELECT *
FROM Patient p
JOIN OrderPatient op ON p.id = op.patient_id
JOIN Order o ON op.order_id = o.id
LEFT JOIN Analysis a ON op.order_id = a.order_id
LEFT JOIN ResultAnalysis ra ON a.id = ra.analysis_id
LEFT JOIN Result r ON ra.result_id = r.id
...
WHERE p.id = target_patient_id
```

The equivalent SQL query is again much longer due to the associative relations instead of the path expressions and, thus, less comprehensible. These two queries also demonstrate the strength of the projection in a graph data model in comparison to the projections in a relational data model. The SQL query returns a table that contains various columns potentially providing a load of information which make it challenging for the user to grasp all the information when viewing the records in the table. The Cypher query, in contrast, returns a subgraph indicating the relationships between the patient and the other entities, such as samples or diagnoses. The subgraph can be inspected more intuitively and lets the user get a better understanding of the relationships that would be hidden in tabular data.

8.5 Conclusion

We proposed a general, intuitive and simple methodology for transforming a relational database schema into graph data schema. The methodology was demonstrated for a medical application (and an implementation based on Neo4j and *NeoDash* – Neo4j Dashboard Builder [5]) for storing, visualizing and analyzing electronic health records, diagnostics and biomedical data. The different transformation steps of the schema graph were shown to obtain the graph database schema from the former relational schema. The benefits of the graph data model like more comprehensible queries and powerful visualisations in comparison to the relational model were also discussed in the context of the tool Graph4Med.

References

[1] Barabási, A.-L., N. Gulbahce and J. Loscalzo. 2011. Network medicine: A network-based approach to human disease. *Nature Reviews Genetics*, 12(1): 56–68.

[2] Girardi, D., S. Wartner, G. Halmerbauer, M. Ehrenmüller, H. Kosorus and S. Dreiseitl. 2016. Using concept hierarchies to improve calculation of patient similarity. *Journal of Biomedical Informatics*, 63: 66–73.

[3] Goh, K.-I., M.E. Cusick, D. Valle, B. Childs, M. Vidal and A.-L. Barabási. 2007. The human disease network. *Proceedings of the National Academy of Sciences*, 104(21): 8685–8690.

[4] Iacobucci, I. and C.G. Mullighan. 2017. Genetic basis of acute lymphoblastic leukemia. *Journal of Clinical Oncology*, 35(9): 975.

[5] de Jong, N. 2022. NeoDash – Neo4j Dashboard Builder. Retrieved 21 Jan 2022, from https://github.com/nielsdejong/neodash

[6] Pai, S. and G.D. Bader. 2018. Patient similarity networks for precision medicine. *Journal of Molecular Biology*, 430(18): 2924–2938.

[7] Edwards, R. 2019. Neomodel documentation. Retrieved 14 Jan 2022, from https://neomodel.readthedocs.io/en/latest/

[8] Rual, J.-F., K. Venkatesan, T. Hao, T. Hirozane-Kishikawa, A. Dricot and N. Li. 2005. Towards a proteome-scale map of the human protein–protein interaction network. *Nature*, 437(7062): 1173–1178.

[9] Tian, X.W. and J.S. Lim. 2015. Interactive naive bayesian network: A new approach of constructing gene-gene interaction network for cancer classification. *Bio-medical Materials and Engineering*, 26(s1): S1929–S1936.

[10] Timón-Reina, S., M. Rincón and R. Martínez-Tomás. 2021. An overview of graph databases and their applications in the biomedical domain. *Database*, 2021.

[11] Wang, B., A.M. Mezlini, F. Demir, M. Fiume, Z. Tu, M. Brudno and A. Goldenberg. 2014. Similarity network fusion for aggregating data types on a genomic scale. *Nature Methods*, 11(3): 333–337.

[12] Wiese, L., C. Wangmo, L. Steuernagel, A.O. Schmitt and M. Gültas. 2018. Construction and visualization of dynamic biological networks: Benchmarking the Neo4j graph database. *In:* International Conference on Data Integration in the Life Sciences (pp. 33–43).

Index